Planning and Installing
Bioenergy Systems
A guide for installers, architects and engineers

Planning and Installing
Bioenergy Systems

A guide for installers, architects and engineers

First published by James & James (Science Publishers) Ltd in the UK and USA in 2005
Reprinted by Earthscan 2005, 2007, 2009

© The German Solar Energy Society (DGS), Ecofys 2005

All rights reserved. No part of this book may be reproduced in any form or by any means electronic or mechanical, including photocopying, recording or by any information storage and retrieval system without permission in writing from the copyright holders and the publisher.

ISBN: 978-1-84407-132-6

Typeset by MapSet Ltd, Gateshead, UK
Cover design by Paul Cooper Design

For a full list of publications please contact

Earthscan Ltd
Dunstan House, 14a St Cross Street
London, EC1N 8XA, UK
Tel: +44 (0)20 7841 1930
Fax: +44 (0)20 7242 1474
Email: earthinfo@earthscan.co.uk
Web: **www.earthscan.co.uk**

A catalogue record for this book is available from the British Library.

Library of Congress Cataloging-in-Publication Data

Planning and installing bioenergy systems : a guide for installers, architects and engineers / German Solar Energy Society (DGS) and Ecofys.
   p. cm.
 Includes bibliographical references and index.
 ISBN 1-84407-132-4 (pbk.)
 1. Biomass energy. I. Deutsche Gesellschaft für Sonnenenergie. II. ECOFYS (Firm)
TP339.P53 2005
333.95'39—dc22

2004006810

This guide has been prepared as part of the GREENPro project co-funded by the European Commission. Also available in the series:
Planning and Installing Photovoltaic Systems: A Guide for Installers, Architects and Engineers
978-1-84407-131-9
Planning and Installing Solar Thermal Systems: A Guide for Installers, Architects and Engineers
978-1-84407-125-8

Neither the authors nor the publisher make any warranty or representation, expressed or implied, with respect to the information contained in this publication, or assume any liability with respect to the use of, or damages resulting from, this information.

At Earthscan we strive to minimize our environmental impacts and carbon footprint through reducing waste, recycling and offsetting our CO_2 emissions, including those created through publication of this book. For more details of our environmental policy, see www.earthscan.co.uk.

This book was printed in Malta by Gutenberg Press.
The paper used is FSC certified and the inks are vegetable based.

Contents

Foreword		**xi**
CHAPTER 1: Introduction		**1**
1.1	The challenge	2
1.2	The universal energy carrier	3
1.3	The potential	4
1.4	The market	4
1.5	The boundary conditions	5
CHAPTER 2: Biomass: energy from the sun		**7**
2.1	How photosynthesis works	7
2.2	Carbon dioxide's key role in climate change	8
2.3	The carbon cycle on our planet	11
2.4	**Biomass as a carbon dioxide store**	12
2.4.1	Growth	13
2.4.2	Energy-efficient resources	13
2.4.3	Long-term use	14
2.5	**Types of biomass**	14
2.6	**Different forms of bioenergy source**	15
2.7	**Utilization of bioenergy sources**	17
2.7.1	Heat	17
2.7.2	Mechanical energy	18
2.7.3	Electricity	18
2.8	**Types of bioenergy source**	19
2.8.1	Solid bioenergy sources	19
2.8.2	Liquid bioenergy sources	22
2.8.3	Gaseous bioenergy sources	23
2.9	**Quality characteristics of bioenergy sources**	24
2.9.1	Solid bioenergy sources	24
2.9.2	Liquid bioenergy	27
2.9.3	Gaseous bioenergy sources	28
2.10	**Solid bioenergy products**	29
2.10.1	Wood pellets	30
2.10.2	Woodchips	31
2.10.3	Logs	33
2.10.4	Wood briquettes	33
2.10.5	Bales of straw	34
2.11	**Liquid bioenergy products**	34
2.12	**Gaseous bioenergy products**	34
2.13	**Possible technical uses**	35
2.13.1	Heat generation	35
2.13.2	Combined heat and power	42

2.13.3	Dimensioning of the combined heat and power system	42
2.13.4	Processing into a product	51

CHAPTER 3: Anaerobic digestion — 53

3.1 Introduction — 53
- 3.1.1 Who should read this chapter? — 53
- 3.2.2 What information will be provided? — 53

3.2 System description and components — 54
- 3.2.1 System description — 54
- 3.2.2 Biogas from manure and co-substrates — 57
- 3.2.3 Various AD systems — 63
- 3.2.4 System components — 66

3.3 Planning an anaerobic digestion project — 77
- 3.3.1 Steps in project development — 77
- 3.3.2 Project creation — 78
- 3.3.3 Feasibility study — 80
- 3.3.4 Project preparation — 87

3.4 Project realization, commissioning and start-up — 88
- 3.4.1 Planning and construction — 88
- 3.4.2 Start-up — 89

3.5 Operation and maintenance — 91
- 3.5.1 Operation of the digester under normal conditions — 91
- 3.5.2 Operation of the digester during malfunction — 92
- 3.5.3 Maintenance — 92

3.6 Economics — 93
- 3.6.1 Introduction — 93
- 3.6.2 Costs — 93
- 3.6.3 Benefits — 99
- 3.6.4 Cost–benefit calculation — 99
- 3.6.5 Example of an anaerobic digester in practice — 100

3.7 References — 100

3.8 Further reading — 100

CHAPTER 4: Liquid biofuels — 101

4.1 Introduction — 101
- 4.1.1 Who is this chapter aimed at? — 101
- 4.1.2 What information is provided in this chapter? — 101

4.2 Biofuels in transportation — 101
- 4.2.1 The market for liquid biofuels — 102
- 4.2.2 The advantages of biofuels — 103
- 4.2.3 Areas of application — 104

4.3 Process for producing liquid biofuels from biomass — 106
- 4.3.1 Natural vegetable oil — 107
- 4.3.2 Biodiesel — 108
- 4.3.3 Ethanol — 109
- 4.3.4 Fuels from synthesis gas — 109
- 4.3.5 Methanol — 111
- 4.3.6 Hydrogen from biomass — 112

4.4 Costs of liquid biofuels — 112

4.5 Liquid biofuels market development — 113
- 4.5.1 Natural vegetable oil — 113
- 4.5.2 Biodiesel — 114
- 4.5.3 Ethanol — 114

4.6	**Using liquid biofuels for mobile applications**	**115**
4.6.1	Natural vegetable oil	115
4.6.2	Biodiesel	115
4.6.3	Ethanol	116
4.7	**Using liquid biofuels for stationary applications**	**116**
4.7.1	The basics	116
4.7.2	Possible technical problems of operating CHP plants with vegetable oil	117
4.8	**Project management**	**117**
4.8.1	General project planning	117
4.9	**Technical planning**	**118**

CHAPTER 5: Small combustion systems — 119

5.1	**Introduction**	**119**
5.1.1	Who should read this chapter?	119
5.1.2	What information is provided in this chapter?	119
5.2	**Heat demand of buildings**	**119**
5.2.1	Detailed measurement of maximum heating output	121
5.2.2	Seasonal distribution of annual heat demand	125
5.3	**Choosing small combustion systems for heating buildings**	**128**
5.3.1	Open fireplaces	130
5.3.2	Closed fireplaces	132
5.3.3	Wood stoves	134
5.3.4	Pellet stoves	138
5.3.5	Central heating cookers	142
5.3.6	Tiled stoves	145
5.3.7	Log-fired central heating boilers	152
5.3.8	Central pellet boilers	158
5.3.9	Woodchip boilers	166
5.3.10	Combination boilers	168
5.4	**Basic design considerations**	**169**
5.4.1	Wood boilers	170
5.4.2	Space heating demand	171
5.4.3	Domestic hot water demand	171
5.4.4	Hot water storage tanks	172
5.4.5	Solar thermal systems	172
5.4.6	Circulation pumps	174
5.4.7	Safety equipment for heating systems	175
5.4.8	Expansion tanks	175
5.4.9	Soundproofing	178
5.5	**Chimneys**	**178**
5.5.1	Chimney flue pipes	180
5.6	**Storage**	**181**
5.6.1	Stores for wood logs	181
5.6.2	Possibilities for storing pellets	182
5.6.3	Storage possibilities for woodchips	197

CHAPTER 6: Large-scale heaters — 201

6.1	**Introduction**	**201**
6.2	**Implementing a wood energy project**	**201**
6.2.1	Seven steps to a successful project	201
6.2.2	Basic conditions for local wood energy projects	203

6.3	**Planning**	**206**
6.3.1	Assessment of the project outline data	206
6.3.2	Assessing the economic efficiency	210
6.3.3	Fuel supply	212
6.4	**Legal organization**	**212**
6.4.1	Options for ownership arrangements	213

CHAPTER 7: Gasification — 217

7.1	**Introduction**	**217**
7.1.1	Who should read this chapter?	217
7.1.2	What information is provided in this chapter?	217
7.2	**System basics**	**217**
7.3	Fundamental principles	218
7.3.1	Gasification	218
7.3.2	Fuel	218
7.3.3	Status of the technology	219
7.4	**Use as energy**	**223**
7.4.1	Gasification applications	223
7.4.2	Possible energy uses of gas generated from wood	224
7.4.3	Combined heat and power in a CHP unit	224
7.5	**Emissions and by-products**	**225**
7.6	**Economic viability**	**226**
7.6.1	Evaluation basis	226
7.6.2	Evaluation of economic viability	227

CHAPTER 8: Legal boundary conditions for bioenergy systems — 231

8.1	**Introduction**	**231**
8.1.1	General legal aspects	231
8.1.2	Erection and operation of bioenergy systems	231
8.1.3	Biomass-related legal issues	232
8.2	**General approvals issues for renewable energy systems**	**232**
8.2.1	Grid access permits	232
8.2.2	Building permits	232
8.2.3	Technical requirements	233
8.3	**The approvals process for bioenergy systems**	**233**
8.3.1	Biomass input	233
8.3.2	Emissions	234
8.3.3	Technology-specific aspects	235
8.3.4	Documents accompanying the approvals process	236
8.4	**Further information**	**236**
8.4.1	UK	236
8.4.2	USA	237
8.4.3	Canada	238
8.4.4	Australia	239
8.4.5	Scandinavia	240

CHAPTER 9: Support measures for bioenergy projects — 243

9.1	**Introduction**	**243**
9.2	**Support mechanisms for renewable energy systems**	**243**
9.2.1	Supporting policies	243
9.2.2	Legislative measures	244

9.2.3	Fiscal incentives	245
9.2.4	Subsidies, grants and loan programmes	245
9.2.5	Administrative support for RES	246
9.2.6	Technology development support	246
9.2.7	Education and information	247
9.3	**General information on financial support**	**247**
9.3.1	Project eligibility	247
9.3.2	Applicant eligibility	248
9.3.3	Essential qualifying (compliance) criteria	248
9.3.4	Application form	248
9.3.5	Type and level of funding	248
9.3.6	Cumulation	249
9.3.7	Actual conditions for support programmes	249
9.4	**Further information on support measures in various countries**	**249**
9.4.1	Sources of information in the UK	249
9.4.2	Sources of information in the USA	251
9.4.3	Sources of information in Canada	253
9.4.4	Sources of information in Australia	254
9.4.5	Sources of information in Scandinavia	255
9.4.6	Sources of information in other English-speaking countries	257
9.4.7	Sources of information at EU level	257
9.4.8	Other sources of information on bioenergy	257

Index   **259**

Foreword

Bioenergy is relied upon worldwide as a modern solution for local energy supply and waste management. Within different sectors, from architecture and engineering to agriculture, a variety of professionals are growing more interested in the application of new technology for generating heat and power from biomass. Large amounts of additional know-how are asked for, in particular about the state of the art of the technology and the actual market situation, and this requires services in planning and design, economics and consultancy.

This book should complement the services described above, and should support the decision-making process in offering up-to-date information on the latest technical developments based on best-practice experience. Finally, this guide should work as an aid for high-quality planning and careful system installation.

Highlights of the guide include the following:

- An overview of bioenergy technologies, including a clear description of the fundamentals of both the technology and its application
- More detailed examination of the different technologies, including anaerobic digestion, bio-fuels, small-scale ovens, large-scale boilers and gasifiers
- Data on the international legal framework and on selected regional, national and international support programmes

Bioenergy is arguably the broadest field in renewable energy, with significant potential for development in each technology subset. By careful selection of material, with a clear description of theory and application including best-practice examples, this guide offers the knowledge and tools for installers to apply this technology in their own context.

1 Introduction

The solar power offered annually in the form of radiation on the earth's surface exceeds the current energy demand of mankind by 11,000 times. Biomass is the stored energy of the sun. Plants convert solar energy with a mean efficiency of 0.1% by photosynthesis and store it lastingly in their parts – leaves, stalks and blossoms. At optimum boundary conditions the energy in the biomass can be stored almost infinitely without losses.

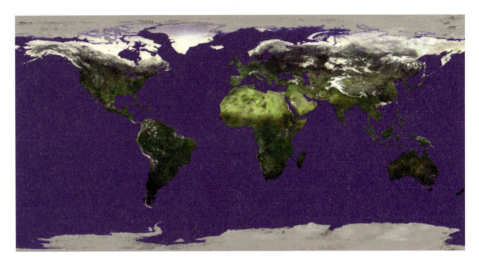

Figure 1.1.
The green planet
Photo: NASA / www.nasa.gov

Biomass is the only renewable energy that can be converted into gaseous, liquid or solid fuels by means of well-known conversion technologies. Accordingly this universal renewable energy carrier can be used in a widespread field of applications in the energy sector. Already today it is possible to provide bioenergy carriers for the entire range of energy-demanding applications, from stationary heat and power supply to the fuelling of mobile applications for transport and traffic.

The annual global production of biomass exceeds today's world's energy consumption by a factor of 13!

The broad range of possible areas for use of biomass as an energy carrier, the advantage of secure and harmless storage, and the possibility of integrating agricultural and forestry enterprises into local energy supply offers a wide, sustainable field of application. The use of biomass as a renewable fuel can reduce the global energy footprint of all nations, and open the door to a sustainable and climate-neutral future.

In contrast to the direct use of solar energy or wind power, biomass as a renewable energy carrier is always available, and can be used to provide transmittable power. Usually, after the biomass has been treated, it is converted to one of these three major forms of energy:

- electricity
- heat
- fuel.

These energy carriers compete with fossil energy carriers in a broad range of applications.

The range of applications and the availability of biomass are only two important advantages of biomass. Another major argument for the use of this energy resource originates from its power regarding climate and environmental protection. When use is made of the stored energy in biomass, greenhouse gases such as carbon dioxide are emitted, but the amount is the same as that produced by natural decomposition processes. Thus bioenergy carriers can be considered neutral as far as the climate-damaging greenhouse effect is concerned.

1.1 The challenge

Energy is the key to the long-term survival of our modern civilization. On average every single human being of the six billion people on earth accounts for 2 tonnes of carbon used for energy purposes each year. But of course there is a large difference between the industrialized and developing countries: for example, a European consumes more than 6 tonnes of carbon – 40 times more of our restricted global energy resources than a human being in Bangladesh.

Figure 1.2.
More than 300 billion tonnes of CO_2 have been emitted to the atmosphere since 1990
Photo: creativ collection/
www.sesolutions.de

Today, 90% of the energy carriers used are of fossil origin, and their use is associated with the emission of carbon dioxide to the atmosphere. Hence every year our earth's atmosphere receives more than 15 billion tonnes of CO_2. Worldwide, scientists agree that proceeding in such a manner will lead to irreversible damage to our climate.

Yet satisfaction of the energy demand of our civilization does not necessarily need to be based on the climate-damaging fossil energy carriers. CO_2-neutral energy resources, such as the direct use of solar energy and wind power and the indirect use of solar radiation in the form of biomass, can provide the necessary energy. A mix of these renewable energy carriers is able to offer all sorts of forms of energy to meet the demands of our modern life.

The European Union (EU) has placed a strong emphasis in its energy policy on the use of bioenergy carriers and the development of a strong bioenergy market. The following ambitious targets were set in the EU's white paper for the EU member countries regarding the use of biomass by the year 2010:

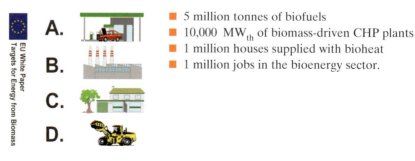

- 5 million tonnes of biofuels
- 10,000 MW_{th} of biomass-driven CHP plants
- 1 million houses supplied with bioheat
- 1 million jobs in the bioenergy sector.

Figure 1.3.
Ambitious targets for bioenergy in the EU until 2010
Graphic: Dobelmann/
www.sesolutions.de

1.2 The universal energy carrier

The use of biomass is the oldest method of supplying energy to mankind. However, modern bioenergy carriers such as wood pellets or chips, logs, wood gas, biogas and plant oil or biodiesel offer interesting potential for providing innovative energy solutions to meet today's energy demand. These natural fuels can be used in stationary applications to provide heat and power to households, to public buildings, in agriculture or in industry. Biodiesel can be used in series production engines for cars, and only minor modifications to car engines are needed for them to be able to drive on plant oil, so that already today mobility and transport problems can be solved fully by using energy crops without polluting the environment and without damaging our climate.

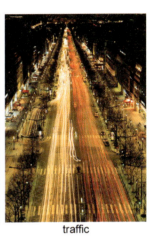

heat   electricity   traffic

Figure 1.4.
Applications of bioenergy
Photo: creativ collection/
www.sesolutions.de

Biomass as a universal and renewable energy carrier is going through a renaissance regarding technology development and reputation. As well as the positive environmental effects of biomass-based energy supply there are several social and economical aspects, as the harvesting, treatment and transport of biomass are labour intensive. With 1.75 new long-term jobs per generated gigawatt-hour of bioenergy comes a significant net job creation – an important criterion for the sustainable development of rural areas both in the EU and in most other countries.

In general, regional areas profit directly from the positive economic effects of using bioenergy. In contrast to an added value of 20% in the fossil-fuel-based energy sector, bioenergy projects provide an added value of 60% to the region.

photosynthesis

Figure 1.5.
Photosynthesis: the natural power plant
Photo: creativ collection/
www.sesolutions.de

Photosynthesis as a natural power plant offers considerable potential to support sustainable structural development and the reinforcement of rural areas in Europe. Hence bioenergy carriers present long-term advantages for rural development, for the security of energy supply, and also for the agricultural production of food, and will improve security of supply in the EU. Biomass as stored solar energy is thus showing its power as a universal element of a sustainable economic policy.

solid   liquid   gaseous

*Figure 1.6.
Types of biomass
Photo: creativ collection/
www.sesolutions.de*

1.3 The potential

On the land areas of our planet grow about 200 billion tonnes of biomass with an energy content of approximately 30,000 EJ (1 EJ = 1 exajoule = 1×10^{18} J). This is equivalent to the energy content of the entire stock of fossil energy carriers on earth. Each year, growth of about 15 billion tonnes of biomass adds an energy potential of 2250 EJ to this amount through photosynthesis.

Unfortunately this vast potential cannot be used directly for energy purposes, as it is spread over the entire landmass of our planet. Only part of this potential is available for the use of biomass as an energy carrier; it is called the technical potential, and has been estimated to be of the order of 150 EJ.

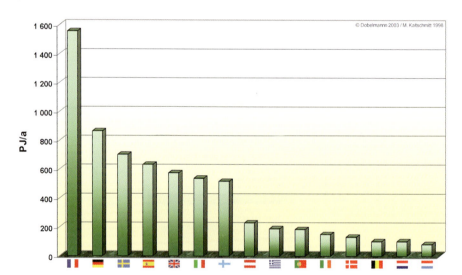

*Figure 1.7.
Technical potential of biomass in Europe
Graphic: Dobelmann/
www.sesolutions.de
Data: M. Kaltschmitt*

The part of the technical biomass potential that can be used in an economically feasible manner depends heavily on the relevant market conditions. Thus local oil and gas prices, and the supporting policy instruments such as subsidies and revenues, complement the environmental and social advantages of bioenergy. But one aspect is clear: with increasing prices for fossil energy carriers, the technical potential for bioenergy projects is enhanced, too.

1.4 The market

Biomass is already making a significant contribution to the security of a sustainable energy supply in a number of European countries.

More than 2200 PJ (1 PJ = 1 petajoule = 10^{15} J) of stored energy in the form of biomass is being harvested in the EU each year; about 1700 PJ of this amount are

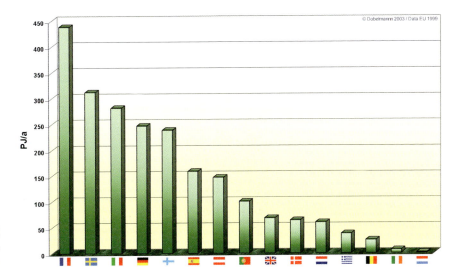

Figure 1.8.
The use of bioenergy in Europe
Graphic: Dobelmann/
www.sesolutions.de
Data: European Commission

used directly to generate heat and 500 PJ are used to generate electricity. The EU has agreed upon a target of an average share of electricity from renewable energy sources of 12% by the year 2010. Biomass alone is expected to provide 10% of the entire European energy supply, equivalent to about 5800 PJ.

Some EU member states are already complying with this target. Finland, followed by Sweden, Austria and Portugal, already provide more than 10% of their energy demand by using biomass. These countries have made use of almost half their biomass potential, and have thus shown that development in the bioenergy sector can lead to sustainable success in this area. This benchmark shows where other countries, such as the two largest EU countries, France and Germany, have to go. France and Germany have developed only about 30% of their existing biomass potential.

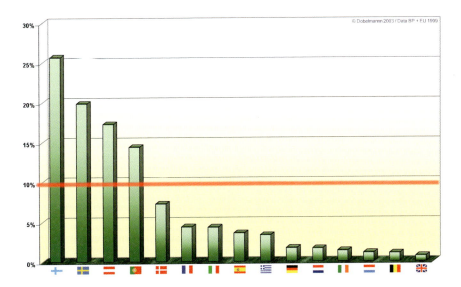

Figure 1.9.
Bioenergy share compared with total
energy consumption
Graphic: Dobelmann/
www.sesolutions.de
Data: European Commission

1.5 The boundary conditions

If we look at the various European countries, we can see that there are wide differences in the boundary conditions for bioenergy projects as far as the administrative and economic aspects are concerned. Even though the use of biomass as an energy carrier to replace fossil energy resources is strongly supported by the EU, administrative hurdles – right down to the local policy level – often hinder the development of bioenergy projects. During the last decade the national feed-in tariffs for bioenergy have been levelling out in Europe. While countries such as Austria, Germany, France and Portugal offer fixed feed-in tariffs for electricity from biomass, other countries, such as the United Kingdom, Italy and Belgium, have introduced

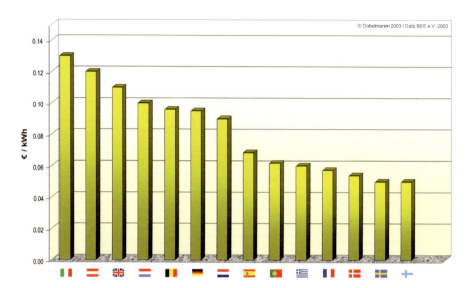

Figure 1.10.
Revenues for electricity from biomass in Europe
Graphic: Dobelmann/
www.sesolutions.de
Data: European Commission

more market-oriented instruments such as renewable energy quotas, together with certificate trading.

A comparison of feed-in tariffs for systems smaller than 2 MWe is shown in Figure 1.10 for the EU member states. Each country commonly offers a number of different classes and divisions, so this figure is meant only as a general guide. Nevertheless, it shows clearly that there is no obvious trend regarding the relation of the type of policy instrument – fixed feed-in tariff or quotas/certificates – and the revenue paid. As can be seen, Italy and Austria – two countries with different policy mechanisms – show the highest revenues for bioenergy in Europe.

In general, the financial revenues for electricity from biomass in each country differ by type, capacity and biomass. In addition, individual investment subsidies complement specific projects together with low-interest loans and tax incentives.

It is often difficult to form a clear overall picture of the bioenergy market and the variety of supporting instruments. In addition, there are frequent changes in the political framework, and investors should therefore check carefully the local site conditions and the regional, national and European support schemes in order to devise financing schemes at the lowest costs and with the minimum risk. The administrative aspects such as permissions procedures also need to be taken into account.

In conclusion, the successful market introduction and increasing penetration of bioenergy carriers always depends on the complex variety of support mechanisms – political, legal, administrative and financial. Countries with a higher share of energy from biomass commonly show long-term targets of bioenergy in their national energy policy, and a bundle of instruments to support the development of bioenergy projects. Technology development, research and education elements also play an important role in the bioenergy sector in these countries. The Finnish bioenergy industry is one of the world market leaders in wood-based bioenergy systems: it is an excellent example of strong support of this sector at government level.

2 Biomass: energy from the sun

Photosynthesis is the ability of plants to build biomass from the carbon dioxide in the atmosphere together with water and nutrients. It is the basis of all life on our planet. There are two key aspects of this essential function in nature.

Plants take their energy from the radiation provided by the sun. This is termed an autotrophic way of living. Plants form the nutritional basis for organisms such as humans and animals, which are heterotrophic lifeforms and are not able to make their own energy from sunlight.

The process of photosynthesis is largely responsible for the release of oxygen that heterotrophic creatures need to breathe.

Figure 2.1.
Symbiosis between plants and mammals
Graphic: Dobelmann/
www.sesolutions.de

The way that plants and animals live on our Earth in symbiosis can be demonstrated in the experiment shown in Figure 2.1. If a lung-breathing mammal – in this case a mouse – is placed in a hermetically sealed vessel, the animal will die within a few minutes through lack of oxygen. This happens because the animal breathes in oxygen and breathes it out again bound in carbon dioxide.

If a plant is present in the same vessel, it absorbs the carbon dioxide that the mammal breathes out, together with sunlight, as it performs photosynthesis. As a waste product it generates oxygen, which the mouse needs to breathe. The mouse will survive, as it is living in symbiosis with the plant.

Our Earth is also a closed system. On the Earth, and in the Earth's atmosphere, essentially the same processes take place. In this case, the forests of our globe supply humankind and the animal kingdom with the oxygen they need in order to live.

2.1 How photosynthesis works

The green leaf pigment chlorophyll is the internal power station in all plants. Powered by the energy from sunlight, plants convert carbon dioxide taken from the air into biomass such as sugar and starch in their chlorophyll-containing leaf cells. As well as solar radiation, for this process the plant requires minerals – as nutrients – and water. These are drawn from the soil through the plant's roots (Figure 2.2).

Looked at in terms of the chemistry involved, the photosynthesis reaction for the formation of sugar takes the following form:

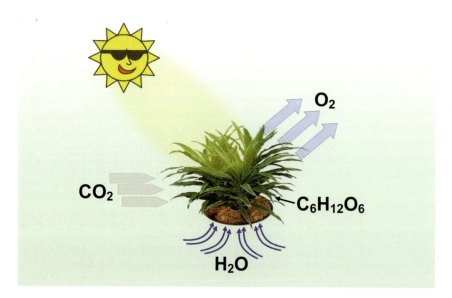

Figure 2.2.
The photosynthetic basis for life
Graphic: Dobelmann/
www.sesolutions.de

$$6CO_2 + 6H_2 \xrightarrow{\text{Sunlight}} C_6H_{12}O_6 + 6O_2$$

Depending on the type of plant, photosynthesis results in the creation of various carbon chains – the carbohydrates. In fast-growing plant species such as maize, photosynthesis in young plants can reach an energy conversion efficiency from sunlight of up to 2%.

Photosynthesis is the only supplier on our planet of the oxygen that humans and animals need in order to live. The particular importance of photosynthesis for all life on Earth becomes clear if we considers the global dimensions of this process and its historical impacts on the Earth's climate.

When our planet was first created, there was no oxygen whatsoever. Our Earth's atmosphere at that time consisted of methane and carbon dioxide. Approximately 3.5×10^9 years ago, photosynthesis emerged as a result of evolution.

It was only as oxygen was released as a result of plants performing photosynthesis that the atmospheric shell began to take on its current composition. This contains around 21% oxygen, for humans and animals to breathe. Each year, plants provide the planet's oxygen-breathing creatures with 1×10^{11} tonnes of fresh oxygen. The element carbon goes through a process of constant change on our planet. Photosynthesis is the crucial factor in this carbon cycle. Each year, global photosynthesis activities take 2×10^{11} tonnes of carbon dioxide out of the atmosphere and store it as usable biomass in plants.

Figure 2.3 illustrates the effects of these processes for a 60-year-old beech tree.

2.2 Carbon dioxide's key role in climate change

Currently, mankind's energy supply is based overwhelmingly on fossil fuels. Every year the power stations and motor cars on our planet burn an amount of fossil biomass that took 500,000 years to grow.

Annually, this results in around 24 billion tonnes of carbon dioxide being released into the atmosphere worldwide. About 80% of this huge amount stems from human activities, such as the burning of mineral oil, natural gas and coal. The natural carbon dioxide content in the Earth's atmosphere is low, but long-term studies on the composition of the atmosphere show that this reservoir has been growing (Figure 2.4).

The combustion of fossil resources such as mineral oil and coal over the last 100 years has resulted in the proportion of carbon dioxide in the Earth's atmosphere increasing by 27%. Whereas in 1765 the CO_2 content in the atmosphere was around 280 ppm (parts per million), today it registers at 360 ppm. Experts ascribe a key role to the globally observed greenhouse effect in this increase.

This tree is around 25m tall and has an approximate crown diameter of 15m². With somewhere in the region of 800,000 leaves it multiplies its active surface area by a factor of ten from 160m² to 1,600m² of leaf surface area. The countless stomata (leaf pores) take 9.4 cubic metres of carbon dioxide out of the air into the leaf cells on a summer's day. The tree processes this, powered by the sun's energy and taking water and nutrients from the soil, into 12kg of carbohydrates (sugars and starches). This process also releases a quantity of 9.4m² of vital oxygen. Just 150m² of leaf area during the growth phase supplies the complete oxygen needs of one person. This tree therefore provides oxygen for 11 people. At the same time it consumes the daily carbon dioxide emissions of two and a half households. If the tree should be felled for any reason and it is desired fully to replace this tree, it would be necessary to plant 2500 saplings each having a crown volume of 1m³. The costs for this planting would run to almost three-quarters of a million pounds.

Figure 2.3.
The value of a tree's services to the environment
Graphic: Adapted from Dobelmann/www.sesolutions.de

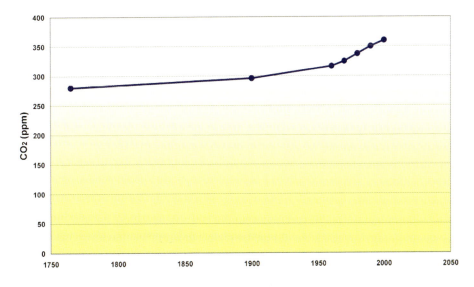

Figure 2.4.
CO_2 content in the atmosphere since 1765
Graph: Dobelmann/www.sesolutions.de.
Data: BINE technical information service

The greenhouse effect describes climatic developments that in the long term will lead to an increase in the average yearly temperatures in the Earth's atmosphere. As well as carbon dioxide, other harmful gases are contributing to this global warming.

The following gases are relevant for the greenhouse effect:

- carbon dioxide (CO_2)
- water vapour (H_2O)
- ozone (O_3)
- methane (CH_4)
- dinitrogen oxide (nitrous oxide) (N_2O).

The enrichment of these gases in the Earth's atmosphere arises essentially from four anthropogenic effects:

- the destruction of the rainforest (15%)
- the use of fossil fuels (50%)
- emissions from the production and application of chemicals (20%)
- emissions from farming (15%).

Figure 2.5 shows how the greenhouse effect works. On average, 342 W/m² (watts per square metre) of short-wave solar radiation reaches the Earth. Of this, around 77 W/m² is reflected straight back into space by the atmosphere, the aerosols in it and the clouds. The Earth's atmosphere absorbs approximately 67 W/m² of the solar radiation.

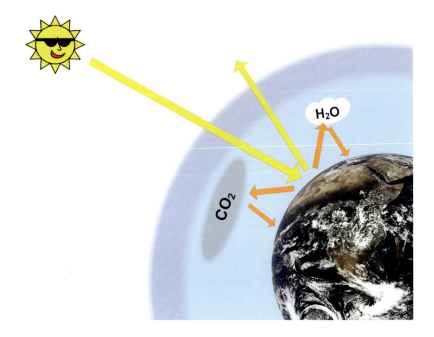

Figure 2.5.
How the greenhouse effect works
Graphic: Dobelmann/
www.sesolutions.de

Hence about 198 W/m² of the incoming solar radiation reaches the Earth's oceans and land masses. The Earth's surface absorbs approximately 168 W/m², and reflects around 30 W/m² directly back into the atmosphere. As it does so, the radiation's spectrum changes from short-wave to long-wave radiation.

Long-wavelength heat rays are not so easily able to penetrate the Earth's atmosphere as the sun's short-wavelength rays. They are reflected back down to the Earth's surface by the atmosphere enriched with carbon dioxide and water vapour, as in a greenhouse. The Earth's surface gives off approximately 390 W/m² of long-wave heat radiation, of which only 66 W/m² stays in the atmosphere; 324 W/m² is immediately reflected back onto the Earth's surface.

Since the beginning of the 19th century, this has already led to an increase in the average global temperature of 0.5°C. According to estimates by climate experts, in the next century this warming may amount to between 1.4°C and 5.6°C.

Even if these temperature figures appear small to the man in the street, the impacts of global warming on this scale will be catastrophic for many countries. A rise in the average annual temperature would mean that many areas of ice on the Earth would melt. As a consequence of the rise in temperature, the sea level could rise permanently by between 11 and 88 cm, threatening the coastal areas in which 50% of the world's population live (Figure 2.6).

Many of the anticipated changes can already be seen. The Arctic ice cap has already shrunk by 10–15% over the last 35 years. In Europe, garden plants flower on average for 10.8 days longer than 35 years ago. Although these may appear to be welcome developments, they are a cause for concern. The increased temperatures also

Figure 2.6.
Storm surge on the sea
Photo: creativ
collection/www.sesolutions.de

Figure 2.7.
Hurricane over the USA
Photo: NASA/www.nasa.gov

result in more frequent extreme weather phenomena, such as storms and floods (Figure 2.7).

The damage caused by weather disasters such as hurricanes has increased more than tenfold since the 1950s. In the 1990s this damage reached record levels of US$40 billion a year.

2.3 The carbon cycle on our planet

On our planet, the element carbon goes through a constant cycle. In this cycle, the carbon is bonded in carbon chains via photosynthesis in plants. Plants serve animals as food. Animals build the flesh of their bodies out of carbon chains. When vegetable or animal biomass rots, the carbon is released as carbon dioxide. The dimensions of this cycle are shown in Figure 2.8. In total, more than 575 billion tonnes of carbon are undergoing constant transformation on our planet.

When biomass is burned, carbon dioxide from the carbon chains is released into the atmosphere. Despite this, scientists do not consider these emissions of carbon dioxide to be relevant for the climate.

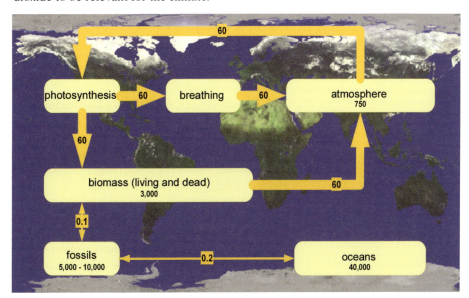

Figure 2.8.
Carbon cycle on Earth
Graphic:
Dobelmann/www.sesolutions.de.
Data: German Bundestag:
Study Commission

The reason for taking this view has to do with the balancing periods that are involved in climatic change. As a tree grows, it absorbs carbon dioxide in its biomass. To create 1 m³ of wood, the tree takes 1 tonne of carbon dioxide out of the atmosphere. Of this, 250 kg are stored as carbon in the wood, and 750 kg are given off as oxygen into the atmosphere.

When the tree dies, a gradual process of decomposition begins in the forest, in which micro-organisms break the carbon chains in the tree down into their constituent parts. Depending on the type of wood and its location, this process can take from one to several years. In the natural cycle of rotting, also termed cold combustion, the same amount of energy is released as was originally locked into the wood through photosynthesis.

In total, this rotting process releases exactly the same amount of carbon dioxide as if the tree were used for energy in a combustion process. The advantage of swift oxidation by combustion is that it creates useful heat.

The routes taken by the processes of combustion and rotting are illustrated in Figure 2.9. As can be seen, the carbon cycle is closed in both cases (combustion and rotting).

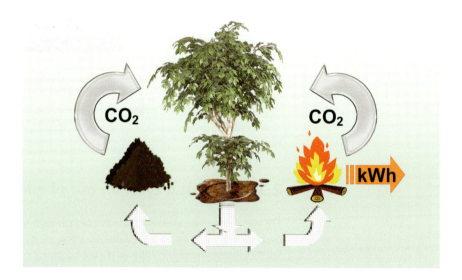

Figure 2.9.
CO_2 cycle of combustion and rotting
Graphic:
Dobelmann/www.sesolutions.de

The release of energy in combustion happens over a period of several hours. It has a high energy density, which can be put to technical uses. Natural rotting takes place over a much longer period, in some cases several years. The resulting energy density is no longer useful to us.

However, for the balancing period of climatic events, this difference does not have any effect. The release of carbon dioxide takes place at a similarly fast speed in relation to the much longer periods of time of the 'balancing periods'. For this reason, the combustion of bioenergy sources is considered to be CO_2-neutral.

2.4 Biomass as a carbon dioxide store

The principles of sustainable forestry and agriculture already offer ways of mitigating the development of a global climate change. Here the continuing expansion of forests and increasing of stocks plays an important part as this increases the carbon dioxide storage capacity of forests.

The main significance of biomass in a stable carbon dioxide balance is as a renewable resource. Active forests and green spaces are carbon dioxide stores that can tie up carbon dioxide taken from the atmosphere for decades at a time. The carbon also stays bonded for a long time after the growth period has ended. Essentially there are three relevant biomass storage mechanisms for carbon dioxide, as listed below.

2.4.1 Growth

New biomass, because it grows quickly, is a particularly active carbon store. In Europe, the amount of stored carbon dioxide is constantly growing. In the forests of Europe, around 793 million m^3 of new wood grow each year, but only around 418 million m^3 are removed from the forests through felling (Figure 2.10).

*Figure 2.10.
A forest in growth
Photo: creativ
collection/www.sesolutions.de*

As a result, the area of forest in Europe has grown by 9 million ha in the last 10 years. Through this increase in their wood reserves, European forests remove about 140 million tonnes of carbon from the air each year. This reduces the amount of the greenhouse gas carbon dioxide in the air and makes an active contribution to climate protection. With active forest management, therefore, it is possible to achieve a lasting reduction in carbon dioxide.

2.4.2 Energy-efficient resources

Products made from biomass materials such as wood and other renewable resources such as fibres and oils can fully replace petrochemically produced fibres and materials.

Unlike resources produced from petrochemicals, renewable resources made from organic materials require much less energy for their extraction, processing and disposal. As a consequence, their production and processing also means lower carbon dioxide emissions into the atmosphere. The emissions of other pollutants are also lower than for mineral-oil-based raw materials.

*Figure 2.11.
Basic chemical materials
Photo: creativ
collection/www.sesolutions.de*

At the end of their life as a consumer product, organic materials can generally be used without restriction for the production of renewable energy. This is in contrast to most chemical products based on fossil sources.

By using wood in such a way, it is possible to save on energy sources such as coal, oil and gas, and to introduce a second lifecycle for the products. The use of the products for energy again closes the natural carbon cycle.

2.4.3 Long-term use

Carbon dioxide storage in biomass lasts beyond the lifetime of the individual tree in the forest. If the wood is harvested from the forests, it is usually processed into building materials and furniture or other long-life economic goods (Figure 2.12). Construction materials are one example of these long-life goods.

Figure 2.12.
Wood is a long-life construction product
Photo: creativ
collection/www.sesolutions.de

The carbon stored in wood and other biological fibre products is retained during the whole lifetime of the product. For example, a tonne of wood used for construction or furniture-making contains 500 kg of carbon, which in turn has bonded 1.8 tonnes of carbon dioxide from the air.

2.5 Types of biomass

Biomass is the total mass of organic substances occurring in a habitat. The forms of biomass on our planet are many and varied (Figure 2.13).

Figure 2.13.
Biomass displaying the diversity of life
Photo: creativ
collection/www.sesolutions.de

Not only are the forms of biomass on our planet varied. There are pronounced differences in the primary uses of biomass by humans. In addition to the food industry, biomass can also be used in engineering industries, such as in the manufacturing of clothing or construction materials.

Figure 2.14.
Biowaste: organic municipal waste
Photo: creativ collection/|
www.sesolutions.de

When the original use ends, this can be followed by a secondary energetic use of biomass. For example, biowaste (organic waste) is a mix of waste materials that can be used as the source material for producing renewable energy (Figure 2.14).

The energy contained in biowaste is generally used via biogas. For this, the waste is generally brought to a tipping site. In the landfill bodies, it is converted by methane bacteria into landfill gas, which contains methane. In some places, however, the direct fermentation of biowaste in anaerobic waste treatment systems can be worthwhile. For biowaste fractions with a high wood content, it is also possible to dry and burn ('combust') the biowaste.

A true cutting-edge field in bioenergy applications is the creation of energy crops that are grown for direct use as a fuel. These will be discussed in the following sections, looking at the various forms in which bioenergy sources appear.

2.6 Different forms of bioenergy source

For a rough overview, biomass can be divided into four categories according to its origin:

- Energy crops. Energy crops are cultivated principally for the generation of energy. Their function, analogous to that of a solar cell, is to capture solar radiation and then store it in their biomass for subsequent use. Examples of energy crops are rape, sunflower, Miscanthus sinensis and maize.
- Post-harvest residues. The residues that occur in the harvesting of cereals and the felling of trees such as straw and forest wood residues are natural waste. This group of 'by-products from nature' is especially suitable for energy recycling because it results in a lowering of production costs for the main products, or can increase the yield of the cultivation chain.
- Organic by-products. The processing of biomass by humans to create products forms a further group of by-products. These by-products include manure and liquid manure from animal husbandry and residues from the industrial processing of wood and vegetable fibres. Here, too, energy recycling can lead to increased profitability and ensure that parts of the production process are permanently environmentally sustainable.
- Organic waste. Organic waste comprises the products that are used by consumers and producers and their residues. This includes domestic biowaste, sewage sludge

and residues from food production, for instance from abattoirs. Organic waste is generally subject to waste legislation. Consequently, a whole range of legal requirements have to be met, from origin legislation to epidemic control.

Vegetable biomass is generally to be found in a solid aggregate state. In addition it has a form and water content that in most cases for technical reasons rule out direct energy use.

It is only possible to create a usable bioenergy product with the correct properties by processing the biomass. In doing this it is also possible to change the aggregate state of the biomass to allow applications such as use as fuel in conventional processes.

Figure 2.15 gives an overview of the most widespread methods of processing the four classes of biomass into bioenergy sources in the familiar aggregate states.

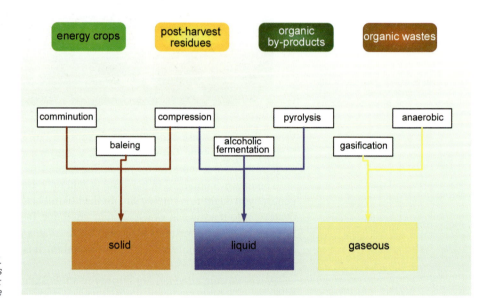

Figure 2.15.
Preparation of biomass
Graphic:
Dobelmann/www.sesolutions.de

For their use as fuels, bioenergy sources are always classified according to the aggregate state they are in: solid, liquid or gaseous (Figure 2.16).

The existing aggregate state essentially determines the possibilities for further use of the bioenergy sources in energy conversion facilities such as engines or combustion plants. Although combustion engines and fuel cells – which are currently still in the research stage – are unable to use solid bioenergy sources, stationary heaters or combined heat and power combustion systems can use solid fuels.

The form and aggregate state of processed bioenergy products are determined by the available conversion technologies and systems. Every utilization facility, such as an

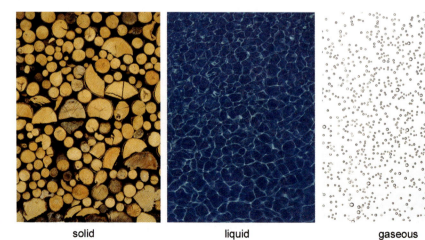

Figure 2.16.
Aggregate states of biomass
Photos: creativ
collection/www.sesolutions.de

engine or a furnace, has a method of operation that is optimized for particular states and qualities of bioenergy sources. For permanent successful operation, these states and qualities have to be maintained within tight limits.

2.7 Utilization of bioenergy sources

There are three fundamental forms of energy in terms of utilization in our modern lives: heat, mechanical energy and electricity (Figure 2.17).

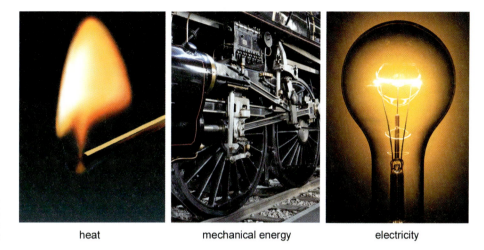

heat   mechanical energy   electricity

Figure 2.17.
Utilization of bioenergy
Photos: creativ
collection/www.sesolutions.de

The use of bioenergy sources can cover all these kinds of energy requirement. Here there are many possibilities for providing the desired energy forms from bioenergy sources with different aggregate states.

2.7.1 Heat

Heat is generated primarily in combustion systems. On a small scale these can heat living space; as large-scale plant they can make their heat available via local heat networks to whole city districts.

For stationary biomass combustion systems that exist solely to generate heat, solid fuels predominate. Wood as a residue or raw material and production residues from farming can be used for heat provision with low processing costs in comminution or drying (Figure 2.18).

Figure 2.18.
Modern wood burner 2 + 3.2 MW
Photos: Schmid
AG/www.holzfeuerung.ch

2.7.2 Mechanical energy

Mechanical energy is required mainly in the transport industry. In practice, it is generated via heat- and power-generating machines such as engines. In these, a liquid or gaseous fuel (available at some filling stations) is ignited in the cylinders of a combustion engine. The expansion of the fuel/air mixture caused by combustion is then converted into power via crankshafts and gears. The heat occurring in this process has to be dissipated to the environment via a cooling system.

The use of biodiesel in Europe (Figure 2.19), the use of a mixture of ethanol in France and the USA, and the use of pure ethanol in Brazil are successful examples of the use of bioenergy sources in the traffic and transport sector.

Figure 2.19
Biodiesel: environmentally friendly mobility
Photo: UfoP/www.ufop.de

With the use of vegetable oils from rape or sunflower seed and alcohol produced from biomass, it is possible to cover our society's mobility needs in an environment- and climate-friendly way. Today, bioenergy fuels are already a technically equivalent alternative to fossil energy sources.

2.7.3 Electricity

The generation of power (electricity) from bioenergy also makes use of the capabilities of heat and power generation. Systems that generate mechanical energy in combustion engines or directly and indirectly fired turbines are coupled to electricity generators. These convert the mechanical energy that is produced into electrical energy, with low losses.

Because the ratio of mechanical energy generated to the production of heat is always approximately one-third power to two-thirds heat, combined heat and power in stationary applications can improve the economic efficiency of these projects as a whole.

Figure 2.20.
Landfill gas utilization with engines
Photo: Dobelmann/www.sesolutions.de

It is primarily biogenic gases that are used for stationary power generation applications. These can be landfill gases, as shown in Figure 2.20. Biogas from the recycling of agricultural or other organic residues is another possibility that is used in many places around the world.

2.8 Types of bioenergy source

Biomass is available on the market in all kinds of different forms. This handbook presents the most important products for the three aggregate states (solid, liquid, gaseous) in their usual marketable forms.

2.8.1 Solid bioenergy sources

The largest group of solid bioenergy sources is products made from wood. These are obtained when firewood is taken from forests and when waste is utilized from the industrial processing of wood products. In many places by-products from agriculture, such as straw, are also used for generating energy from biomass.

When forests are thinned, apart from the trunk of the tree, which is used for the furniture and construction industries, wood residues of a lesser quality are also collected. For each hectare of forest, between 0.4 and 0.8 tonnes of air dry firewood can be obtained from these forest wood residues. Together with the other quantities of wood residues that are produced during forest maintenance, this gives an annual fuel yield from a permanently used forest area of around 1.5 t/ha.

Figure 2.21.
Mechanised wood harvest with a harvester machine
Photo: Zeppelin AG/www.zeppelin.de

In modern wood harvesting, the trees are felled by harvester machines (Figure 2.21). These cut down the tree using a gripper arm with a chainsaw mounted on it. In addition, harvester machines can automatically remove the branches from the trunk, strip off the dark wood bark, and cut the trunk into transportable lengths. This means that part of the value-increasing wood processing is carried out before the wood leaves the forest.

When the round trunks are machined into planks and beams, large amounts of residues are produced. However, for the most part these are utilized in the wood industry for other materials. Woodchips and shavings that are free of any bark, for example, are the base products for high-value chipboard sheets (Figure 2.22).

Figure 2.22.
Industrial wood and its by-products
Photo: Dobelmann/www.sesolutions.de

However, another part of these residues still has fragments of dark bark attached and is therefore unsuitable for utilization as a wood product (Figure 2.23). These bark pieces are ideal for energy recycling. Because of the high ash content, these residues are utilized mainly in larger heat supply stations and combined heat and power plants as a co-firing substrate.

Figure 2.23.
Bark: by-product of wood processing
Photo: Dobelmann/www.sesolutions.de

Other significant residues from agriculture include straw (Figure 2.24) and other stem products such as hay. These post-harvest residues are often available locally in large utilizable quantities.

Figure 2.24.
Mechanised straw harvest with bale press
Photo: Claas AG/www.claas.de

Figure 2.25.
Straw is a natural residual product
Photo: creativ
collection/www.sesolutions.de

The straw from 1 ha of cereal has an energy content of 73 GJ. This is roughly equivalent to 2000 litres of heating oil. However, straw and other stem products have combustion characteristics that are different from those of ligneous fuels. The ash melting point and emission behaviour of straw-type biomass mean that different technical approaches to utilization are required.

To date, it has been possible to achieve large-scale energy recycling of stem-product fuels in cogeneration plants, but not in decentralized installations.

Yet it is not only residue materials produced directly with the creation of biomass that can be considered for energy utilization. Products at the end of their lifecycle that are no longer useful as a material asset are ideal for energy recycling. The processing and combustion of old wood is one example of how this kind of bioenergy product can be obtained from secondary raw materials (Figure 2.26).

Figure 2.26.
Industrial wood waste processing
Photo: Dobelmann/www.sesolutions.de

Because of its previous use, this product can be contaminated with foreign matter such as chemicals, paints or similar. For this reason, many countries have restrictions on the energy recycling of old wood. Burning the wood in small combustion systems is often allowed only if the wood processing has been purely mechanical and the wood contains only insignificant contaminants (Figure 2.27).

Another important category of residues that is not necessarily part of the old wood sector is wood residues from landscape management (Figure 2.28). These occur during maintenance works by roads and waterways, and through work in parks. Wood residues from landscape management are usually a mix of wood, leaves and stem products. Only very rarely would the utilization of this mixture in a new product be considered.

Figure 2.27.
Mechanically prepared wood
Photo: Dobelmann/www.sesolutions.de

Figure 2.28.
Residues from landscape management
Photo: Dobelmann/www.sesolutions.de

Energy utilization suggests itself as a means of disposing of these materials. The combustible quality of these wood residue mixes can be classed as low, owing to the large number of impurities. Because of the soil that is generally still attached, these materials have a high ash content. The other visible impurities such as plastic wrappers, bags and other man-made garbage lead to high levels of toxic matter, with the result that the law demands the controlled disposal of the ash.

2.8.2 Liquid bioenergy sources

Mobility is central to our modern industrialized society. Apart from a few exceptions, the transport of people and goods is sustained by liquid fuels. Today there are already various technically equivalent liquid bioenergy sources available that can take over these tasks. Ethanol from alcoholic fermentation and methanol from lignocellulose biomass such as wood are biogenic fuels. By far the most widespread energy crops, however, are rape and sunflower, the oil from which is used either in its naturally occurring form or as biodiesel (Figure 2.29).

Figure 2.29.
Oil crops found throughout Europe
Photo: UfoP/www.ufop.de

As well as being CO_2-neutral, fuels from liquid bioenergy sources also have better emission properties and a lower potential for pollution. The fact that they nevertheless rank behind fossil fuels when it comes to performance has been demonstrated in various different trials in motor sports (Figure 2.30).

Figure 2.30.
Biodiesel in motor sports
Photo: UfoP/www.ufop.de

2.8.3 Gaseous bioenergy sources

Gaseous bioenergy sources are the result of converting natural biomass. They can be produced by microbiological processes, such as anaerobic methane fermentation, but they can also arise through the thermochemical conversion of solid biomass in gasification processes.

Biogas is created in the fermentation of vegetable and animal biomass without the action of oxygen. Here a symbiosis of bacteria groups brings about the breakdown of carbon compounds into the gaseous end-products methane (CH_4) and carbon dioxide (CO_2). In practice, this happens in agricultural biogas installations or in landfill bodies for example (Figure 2.31).

Figure 2.31.
Gas well at a landfill site
Photo: Dobelmann/www.sesolutions.de

Thermochemical conversion of solid bioenergy sources into combustible gases takes place during gasification, or during smouldering with an oxygen deficit. From the carbon chains in the biomass this creates the combustible gases carbon monoxide (CO), hydrogen (H_2) and, in small amounts, methane (CH_4).

Figure 2.32.
Thermochemically produced wood gas
Photo: Dobelmann/www.sesolutions.de

2.9 Quality characteristics of bioenergy sources

2.9.1 Solid bioenergy sources

There are various ways of classifying solid bioenergy sources. The most important quality feature, as for any energy source, is the calorific value. This is influenced directly by the water content of the bioenergy source.

The lower calorific value, LCV, of wood can be calculated using the following formula:

$$\text{LCV} = \frac{\text{CV}_{wf}(100-w) - 2.44w}{100}$$

where CV_{wf} is the calorific value of water-free wood, and w the water content of the wood in its natural state.

Drawn as a graph, the relationship is as shown in Figure 2.33.

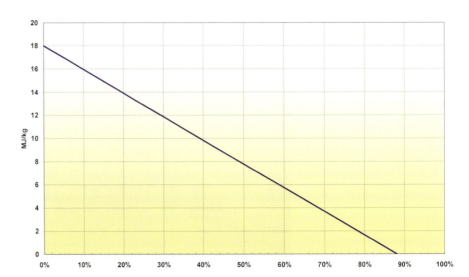

Figure 2.33.
Water content and calorific
value of biomass
Graph: Dobelmann/www.sesolutions.de

Biomass is a natural product. Its natural water content therefore fluctuates considerably, even if it is not artificially changed by external influences such as rain or air humidity. In operational practice, the quickest way to perform calculations is with values based on many years' experience.

The typical water content for fresh ligneous biomass is between 40% and 60%. Green plants can have an even higher water content of up to 80%. If the biomass is dried in the air, a natural water content is reached that, depending on the season and the ambient humidity, varies between 12% and 18%.

Artificially dried bioenergy products such as wood pellets and briquettes have a maximum water content of 10%. However, improper storage can lead to wood pellets absorbing water. A water content above 10% makes pellets unusable.

Because of the large influence of water on the weight, bioenergy sources are handled primarily in volume measurements. Therefore, for determining the volume-based calorific value of solid bioenergy sources, which is important in operational practice, the storage method and physical shape are very important.

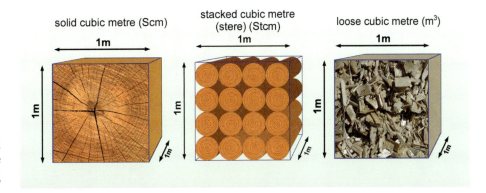

Figure 2.34. Measurement units for timber merchants Graphic: Dobelmann/www.sesolutions.de

For wood there are three main cubic measures, based on the cubic metre as the measurement unit: see Figure 2.34. Because of the different storage densities of the wood, these measurement units result in different weights and volumes of wood. Table 2.1 allows conversion of these historically developed units.

Table 2.1. Conversion table for wood measurement units

	Scm	Stcm	m³
Scm	1	1.43	2.43
Stcm	0.70	1	1.70
m³	0.41	0.59	1

The solid cubic metre calculation unit (Scm) is used only for solid wood. Stacked cubic metres, or steres (Stcm), are used mainly for stackable wood such as metre pieces or logs. The loose cubic metre (m^3) measurement is illustrated above for wood chips, but it is also used for wood products such as pellets, sawdust and shavings, as well as for cereals and other bulk goods.

Table 2.2 shows typical characteristic values for solid wooden bioenergy sources.

In a water-free state, which can only be reached through artificial drying, mature wood has a calorific value of 18.5 MJ/kg.

The ash residues that occur when the wood is burned have a high content of plant nutrients such as calcium, magnesium, potassium and phosphorus. Coarse ash residues (bulk density over 900 kg/m^3) have a low heavy metal content. These are therefore usually allowed to be applied as plant fertilizers.

However, during the combustion of industrial wood residues in systems with combustion capacities of over 150 kW, large amounts of fine ash (bulk density less than 400 kg/m^3) can occur. Occasionally these can contain such high concentrations of heavy metals that, for environmental reasons, their use as fertilizers is not justifiable.

Solid bioenergy sources made from stem products have the characteristic values listed in Table 2.3. With stem products, the ash content and its melting behaviour under the influence of temperature play an important role. In contrast to wood, the ashes from stem products start melting in temperature zones as low as 710–930°C.

Table 2.2.
Characteristic data for solid fuels made from wood
Data: Basisdaten Bioenergie

Wood product		Mass (kg)	Water content (%)	Calorific value (MJ/kg)	Energy content (kWh)	Heating oil equivalent (l)	Ash content (kg)
Measured weight 1 t (solid wood)							
Hardwood	Air dried	1000	18	14.6	4069	407	4.1
(beech)	Naturally dried	1000	35	11.1	3085	308	3.3
	Green	1000	50	7.9	2212	219	2.5
Softwood	Air dried	1000	18	14.9	4137	414	4.9
(spruce)	Naturally dried	1000	35	11.3	3139	314	3.9
	Green	1000	50	8.1	2315	225	3.0
Measured weight 1 t (wood products)							
Wood pellets	Kiln dried	1000	10	17.0	4725	471	5.3
Sawdust	Kiln dried	1000	10	17.0	4536	453	5.4
Shavings	Kiln dried	1000	10	17.0	4425	442	5.8
Loose measurements 1 m³ (woodchips)							
Hardwood	Air dried	283	18	14.6	1161	115	1.2
(beech)	Naturally dried	375	35	11.1	1050	108	1.2
	Green	464	50	7.9	1028	103	1.2
Softwood	Air dried	202	18	14.9	838	84	1.0
(spruce)	Naturally dried	265	35	11.3	792	81	1.0
	Green	332	50	8.1	750	75	1.0
Loose measurements 1 m³ (wood products)							
Wood pellets	Kiln dried	600	10	17.0	2835	283	3.2
Sawdust	Kiln dried	202	10	17.0	823	82	1.1
Shavings	Kiln dried	120	10	17.0	580	58	0.9
Stacked cubic measurements 1 Stcm (split logs)							
Hardwood	Air dried	482	18	14.6	1961	196	2.0
(beech)	Naturally dried	608	35	11.1	1875	188	2.0
	Green	669	50	7.9	1796	181	1.9
Wood pellets	Kiln dried	345	18	14.9	1429	143	1.7
Sawdust	Kiln dried	436	35	11.3	1368	137	1.7
Shavings	Kiln dried	517	50	8.1	1305	131	1.6

Table 2.3.
Characteristic data for fuels made from stem products
Data: Basisdaten Bioenergie

Stem product		Mass (kg)	Water content (%)	Calorific value (MJ/kg)	Energy content (kWh)	Heating oil equivalent (l)	Ash content (kg)
Measured weight 1 t							
Wheat straw	Naturally dried	1000	15	14.4	4032	403	57.0
Barley straw	Naturally dried	1000	15	14.7	4116	412	48.0
Rye straw	Naturally dried	1000	15	14.7	4116	412	48.0
Rape straw	Naturally dried	1000	15	14.3	4004	400	62.0
Maize straw	Naturally dried	1000	15	14.8	4144	414	67.0
Meadow hay	Naturally dried	1000	15	14.3	4004	400	71.0
Hemp straw	Naturally dried	1000	15	14.2	3976	398	27.0
Miscanthus	Naturally dried	1000	15	14.9	4172	417	39.0
Wheat grains	Naturally dried	1000	15	14.2	3976	398	39.0
Stacked cubic measurements 1 Stcm (bale storage)							
Wheat straw	Naturally dried	135	15	14.4	544	54	7.7
Barley straw	Naturally dried	133	15	14.3	533	53	7.6
Rye straw	Naturally dried	140	15	14.9	584	58	8.0
Rape straw	Naturally dried	133	15	14.3	533	53	7.6
Maize straw	Naturally dried	139	15	14.8	576	58	7.9
Meadow hay	Naturally dried	133	15	14.3	533	53	7.6
Hemp straw	Naturally dried	131	15	14.2	521	52	7.5
Miscanthus	Naturally dried	140	15	14.9	584	58	8.0
Loose measurements 1 m³							
Wheat grains	Naturally dried	760	15	14.2	3022	302	43.3

Temperatures on this scale are quickly reached during combustion. Combustion systems for stem goods therefore need to be designed specially to prevent the occurrence of slags or caking of ash onto the furnace grates or walls. The walls and grates of these systems often have a water-cooled design for this purpose.

Stem products have an average chlorine content of 0.5%. Because of the high chlorine and potassium content of straw-type bioenergy sources, they have a high corrosive potential.

Chlorine, like the other natural component substances sulphur and nitrogen, is present in quantities that constitute a relevant emission factor. This results in legal requirements for the operation of installations that burn straw products, which can turn out to be a technical challenge to meet in practice. This applies both to air emissions and to the subsequent utilization of the ash residues, which have a bulk density of 150 kg/m^3.

2.9.2 Liquid bioenergy

Of the liquid bioenergy sources, currently vegetable oil, biodiesel and ethanol have commercial applications in the market.

2.9.2.1 NATURAL VEGETABLE OIL

The use of natural vegetable oil in combustion engines is so new that the European standardization committees have not yet made any final decisions. Until a definitive standard emerges for the use of vegetable oil in engines, a quality standard has been defined by a number of research institutes (Table 2.4).

Table 2.4. Characteristic data for vegetable oil. Data: Ölmühle Leer/www.biodiesel.de

		Vegetable oil German RK quality standard
Density at 15°C	g/ml	900–930
Flash point	°C	220
Max. water content	ppm	750
Max. kinematic viscosity	mm^3/s	35
Acid value	mgKOH/g	2
Max. carbon residue	%	0.4
Min. oxidation stability	h at 110°C	5
Max. phosphor content	ppm	15
Ash content	%	0.01
Max. dirt content	mg/kg	25

Reports exist for vegetable oils that meet these quality standards, showing successful applications as fuel in adapted diesel engines.

2.9.2.2 BIODIESEL

The most important quality characteristics for fatty acid methylester (biodiesel) are regulated, for Europe, in the draft standard pr EN 14 214 (Table 2.5).

Table 2.5. Characteristic data for biodiesel. Data: Ölmühle Leer/www.biodiesel.de

		Diesel-K FAME pr EN 14 214
Density at 15°C	kg/m^3	875–890
Flash point	°C	100
Max. water content	ppm	300
Max. kinematic viscosity	mm^3/s	3.5–5.0
Acid value	mgKOH/g	0.5
Total glycerine	%	0.25
Free glycerine	%	0.02
Max. phosphor content	ppm	10
Methanol content	%	0.3
Temperature steps	°C	−20, −10, 0

Only fuels that meet these criteria are approved for use in production vehicles. If a fuel is used that deviates from this standard, this will automatically lead to the loss of manufacturer's approval and any warranty entitlement relating to it.

2.9.2.3 ETHANOL

Ethanol is used only to a limited extent as a pure fuel for powering petrol engines. It is used mainly as a blending component with fossil fuels. It is possible to mix up to 10% by volume of ethanol with fuels for petrol engines without the need to convert the engines.

Ethanol from biomass, meeting the quality requirements set out in Table 2.6, is suitable for unrestricted mixing with fossil fuels. In the production of this fuel mixture, because ethanol is readily water soluble, it must be ensured that there is no contamination with water when filling or in storage. Hence the production and filling of this fuel mixture generally only takes place in the large regional depots and distribution centres of the fuel manufacturers.

Table 2.6.
Characteristic data for ethanol
Data: Williams/
www.williamsenergypartners.com

Ethanol purity	%	98
Other alcohols	%	< 0.5
Max. water content	%	0.82
Max. dirt content	mg/l	50
Chloride content	mg/l	32
Copper content	mg/l	0.08
Min. pH value	–	6.5
Max. pH value	–	9
Max. acetate content	ppm	7
Visible dirt particles	–	None

2.9.3 Gaseous bioenergy sources

Gaseous bioenergy sources are obtained from the conversion of predominantly solid biomass or from animal wastes such as manure. Anaerobic methane fermentation and thermochemical production of synthesis gases are two different methods of transforming biomass into a gaseous bioenergy source.

2.9.3.1 BIOGAS

The main quality characteristic of biogas is the methane content. Methane has a calorific value of 39.8 MJ/standard m^3 (a m^3 with defined temperature and pressure conditions) and, as the predominant combustible component, determines the energy content of the biogas.

The methane content fluctuates with the mass distribution in the contents of the fermented substrates of carbohydrates, fats and proteins. On average the methane content is around 50–75% by volume, complemented by 50–25% by volume of carbon dioxide.

In addition, biogas can contain small amounts of other gases such as hydrogen sulphide (H_2S), oxygen (O_2) and hydrogen (H_2). Oxygen and hydrogen are not a

Figure 2.35.
Modern biogas facility with
co-fermentation
Photo: Loick Bioenergie/www.enr.de

problem for subsequent energy utilization, but hydrogen sulphide is a harmful gas. As well as being toxic, this gas is also highly corrosive. With H_2S contents above 50 ppm the desulphurization of these biogases is recommended so that subsequent utilization facilities do not find themselves facing increased maintenance costs due to corrosion damage.

2.9.3.2 SYNTHESIS GASES

Synthesis gases are produced during the gasification and smouldering of biomass under oxygen deficit or exclusion conditions. The combustible components of these gases consist of hydrogen (H_2), carbon monoxide (CO) and methane (CH_4). Inert components in these gases are carbon dioxide (CO_2) and nitrogen (N_2).

The relative composition of the gas components depends on the oxidant chosen to initiate the synthesis process. If ambient air is used, this results in a nitrogen component of around 50% and a lean gas is produced (approximately 5 MJ/standard m^3). If pure oxygen is used as the oxidant, this results in a synthesis gas with a high content of hydrogen and carbon monoxide and hence a calorific value of over 10 MJ/standard m^3.

As well as the chemical composition of the gas, particulate dust loads and tar contents are especially important in determining the subsequent possible uses of the synthesis gases. For use in engines, both parameters – dust and tar – must be below 100 mg/m^3, otherwise the long-term operation of engines is not possible.

2.10 Solid bioenergy products

There are many solid bioenergy products on the market that can be used in combustion or gasification systems. The main sources of these energy products are forestry and agriculture and their by-products, and the secondary raw materials sector.

The main sources and end products of solid bioenergy products based on wood that are available on the market are shown in Figure 2.36.

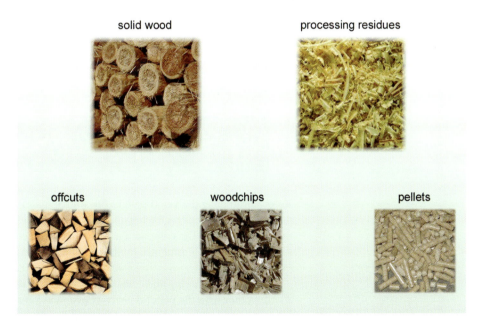

Figure 2.36.
Bioenergy products from wood
Graphic:
Dobelmann/www.sesolutions.de

Modern boilers and combustion systems are optimized in their combustion zones for particular shapes and sizes of energy source. Manually fed wood-burning boilers can only take certain lengths of split logs, and material that is too fine does not burn in the optimum way. Automatically fed pellet or woodchip burners can also be operated only with particular fuel geometries and water contents.

However, it is not just the geometry that is decisive for trouble-free operation of biomass combustion systems. The chemical composition of the fuels also plays an important role in clean combustion. An example of this is grains of wheat and wood pellets. Both have an almost identical density (grains of wheat 750 kg/m^3 and wood pellets 650 kg/m^3) and similar geometry. Yet the different ash melting behaviour of these products (grains of wheat approximately 800°C and wood pellets over 1500°C) means that wheat grains can be used only in special burner systems with water-cooled grates.

The following sections present the most important bioenergy sources available on the market.

2.10.1 Wood pellets

Pellets are used in many sectors to turn powdery bulk material into a mechanically stable form. Pelleting provides many process engineering advantages, such as a high product density and favourable flow and feeding properties. Pelleted products can be transported dust-free using existing conveyor systems such as screw conveyors or suction equipment.

The advantage of using wood pellets is in their standardised size. This allows the manufacturers of wood boilers, even at the low output range of up to 50 kW, to implement fully automatic heating using the solid fuel wood. Wood pellets are dimensionally stable, and their high energy density enables users to have similar delivery intervals for their heating supplies as would be the case for the fossil energy source heating oil.

Wood pellets for use in heating consist of natural sawdust or uncontaminated wood shavings (Figure 2.37). During manufacture, 6–8 m^3 of wood shavings or sawdust are compressed under high pressure into 1 m^3 of wood pellets.

Figure 2.37.
Wood shavings from a planer
Photo: creativ
collection/www.sesolutions.de

Before producing the wood pellets, the source materials are dried. The production process results in pellets having a length of between 5 mm and 45 mm, depending on the manufacturer. The compaction at over 100 kPa enables the wood pellets to remain stable even under mechanical loads during transport and filling, until they come to be burned.

As a result of drying and compaction, pellets have a maximum water content of 8%. Together with the pellet density of over 650 kg/m^3, wood pellets achieve constant calorific values of between 4.9 and 5.4 kWh/kg. A practical rule of thumb states that 2 kg of wood pellets replaces about 1 litre of heating oil.

In the manufacture of wood pellets (Figure 2.38), other natural bonding agents such as maize starch are sometimes used in addition to wood. They are added to facilitate the pressing process. They improve the energy balance and the abrasion resistance of the product. The upper limit for bonding agents is 2%, in order to minimize the ash content and ensure that the ash matrix is optimum for machine handling.

Even if the production of wood pellets with its pressing and drying processes is energy intensive, the energy requirements are well below 2% of the energy content of

Figure 2.38.
Freshly pressed wood pellets
Photo: Dobelmann/www.sesolutions.de

the end product. At this rate, wood pellets are significantly better than fossil energy sources, for which 10–12% of their own energy content is required for their refinement.

Figure 2.39.
Industrial wood pellet production
Photo: Umdasch
AG/www.umdasch.com

2.10.2 Woodchips

For automated heating with wood in systems with higher output ranges – in excess of 50 kW – wood chippings are generally used. These are produced using mechanical choppers from the residues of wood harvesting and wood processing.

For the production of woodchips, there are three different mechanical chopper devices available: the slicer, the drum chopper (Figure 2.40) and the screw chopper.

The specific energy requirements of the chopping process vary between 2 kWh and 5 kWh per tonne of chopped product. This is less than 0.5% of the energy contained in the wood. This requirement is strongly dependent on the water content of the wood. Hard, air-dry wood requires around 18% more energy for the chopping process than moist wood fresh from the forest.

Woodchips are generally between 1 cm and 10 cm long. They are up to 4 cm wide, and are divided into three commercial categories: fine chopped < 3 cm, medium chopped < 5 cm, and coarse chopped < 10 cm. Optimum uniformity of size of woodchips and a low water content are the requirements for problem-free use in automated heating systems.

There must not be any impurities such as stones, metal objects or other foreign matter among the woodchips. Also, for clean combustion the woodchips must not be too moist. Wood fresh from the forest has an average moisture content of 50%. Water content at this level is enough to cause technical problems with combustion.

Automated combustion systems can ensure a smooth sequence in fuel logistics only if the woodchips have the same edge lengths and there are no overlengths in the chopped material; otherwise blockages and bridge formation in the stock can occur in the conveyor systems, causing a stoppage in the combustion system.

Commercially available woodchips are usually pre-dried slightly when sold, as a result of transport and storage. If their water content is below 40%, they are classed as moist and require further drying. A water content of around 20%, termed air-dry, can only be reached after several weeks of drying.

High-quality woodchips for use in automated combustion systems (Figure 2.41) contain no or minimal amounts of bark. This makes it possible to ensure that optimum combustion takes place with a minimum ash content of less than 0.5%.

Figure 2.40.
Drum chopper for 100 m³ of woodchips per hour
Photo: Dobelmann/www.sesolutions.de

As for wood pellets, only pure wood may be used for woodchips. Impurities in the form of plastics or paints, which cannot be ruled out if old wood is used, lead to increased pollutant emissions and ash contents. For this reason, their use in wood boilers without exhaust gas purification is generally prohibited.

Figure 2.41.
High-quality wood chippings
Photo: Dobelmann/www.sesolutions.de

Wood processed into woodchips can be used in virtually all available combustion systems (including shaft furnaces, underfeed and grate-stoker furnaces, and fluidized-bed boiler units). The full required output range, starting at a thermal output of 50 kW and going up to several tens of megawatts, can be covered with this bioenergy product in a fully automated combustion system.

2.10.3 Logs

The production of split logs is the traditional form of wood preparation for energy purposes. For this, the wood is sawed into pieces no more than 1 metre long. Fuel merchants have also established three further log lengths: 25 cm, 33 cm and 50 cm. All boiler manufacturers have optimized their combustion chamber geometries for these.

After being cut to the desired length, the wood is split to optimize the surface area for combustion and to facilitate drying of the wood. When the wood is split manually, the log is stood on end and split into four parts lengthways using a splitting axe or a splitting maul. It can also be done by a machine with hydraulic wood splitters (Figure 2.42). These are able to create 3–5 steres of split logs in an hour.

Figure 2.42.
Industrial wood splitting
Photo: Biomassehof Allgäu
GmbH/www.holzbrennstoffe.de

It is important for the use of split logs in heaters that only healthy, dry wood is used. Good log wood has a water content of less than 20%. This is reached after two years' storage in outside air. If these requirements are met, a residual ash content of the logs of less than 0.5% can be expected.

The production of logs is the most energy-efficient way of preparing wood as a bioenergy product. Mechanical log splitting requires well under 0.1% of the total energy content. Because of their great lack of uniformity, logs are not suitable for automated combustion. Their use is restricted to manually fed solid wood boilers.

2.10.4 Wood briquettes

A fuel variant that is geometrically similar to split logs is wood briquettes. These are pressed in the same way as wood pellets, from wood shavings and sawdust. Here too, only bark-free wood is used. During the manufacturing process the wood is dried to a water content of no more than 10%.

Like pellets, wood briquettes are more densely stored than split logs (Figure 2.43). This is achieved through high compaction in the manufacturing process, compressing one stere of hardwood into 450 kg of wood briquettes. With a calorific value of 18.5 MJ/kg, wood briquettes achieve calorific values that are almost identical to those of brown coal briquettes.

A high energy density, the good heat properties of the compact wood material and the low residues with a maximum 0.5% ash content make wood briquettes an ideal fuel for manually fed small combustion systems such as wood-burning stoves and tiled stoves. Because they do not contain any tree resins, they are not prone to spitting sparks. For this reason, wood briquettes are also highly suitable for use in open fireplaces.

*Figure 2.43.
Wood briquettes in storage
Photo: Umdasch
AG/www.umdasch.com*

2.10.5 Bales of straw

Straw and other stem products are handled in compacted and strapped bales. Rectangular bales have length dimensions of 80–250 cm, widths of 30–120 cm and heights of 30–130 cm. The compaction during the production of these bales achieves storage densities of 130–160 kg/m^3.

Round bales, depending on the machinery available, can be produced with diameters of 60–180 cm. These bales have a length of between 120 cm and 150 cm and achieve storage densities up to 120 kg/m^3.

2.11 Liquid bioenergy products

Biodiesel is the only liquid bioenergy product in Europe that is offered on the general market at filling stations. Although the large mineral oil companies throughout Europe do not offer biodiesel on their forecourts, in many countries there is a dense network of independent filling stations that do supply the biodiesel product (Figure 2.44).

*Figure 2.44.
The biodiesel logistics path
Photos: UfoP/www.ufop.de*

Ethanol and other liquid bioenergy sources are used primarily as blending components and additives in Europe and the USA. As a result the customer is generally unaware of using them.

The direct use of pure ethanol as a fuel, as was implemented on a large scale in Brazil, is an exception on the world market.

2.12 Gaseous bioenergy products

Gaseous bioenergy products are generally produced for stationary applications. Even if biogas is fed into natural gas grids in some places, gaseous bioenergy products are generally tied to one installation and not freely available on the market. In isolated cases the use of biogas for powering automobiles and farm machinery has also been tried out.

2.13 Possible technical uses

As was shown in section 2.7, bioenergy sources are used in three main application fields in the energy industry:

- pure heat production
- electricity generation, and combined heat and power
- vehicle fuel.

In all areas, bioenergy sources can fully replace fossil energy sources such as coal, mineral oil and natural gas.

The main application area for solid bioenergy sources is in the generation of heat. This can be generated efficiently in combustion systems of small (from 3 kW) and medium size (around 100 kW), and in large heat supply stations (around 10 MW) with coupled local heat networks. Liquid and gaseous bioenergy sources are rarely used for pure heat production. Their application area tends to be more in electricity generation or for use as a vehicle fuel.

The processes used to generate electricity from biomass have for many years been part of the best available technology in electrical energy generation. These range from mini power stations with engine-based utilization through to large power stations with steam turbines.

Figure 2.45 shows the most important process methods of power generation and the relevant technologies that they use.

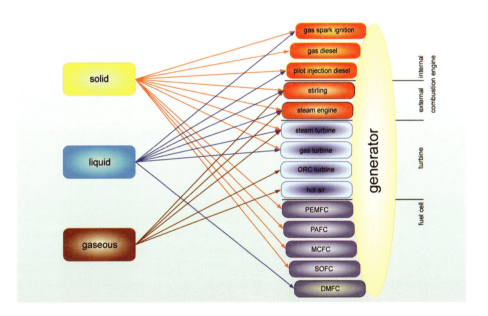

Figure 2.45.
Bioenergy sources in power generation
Graphic:
Dobelmann/www.sesolutions.de

The final application area of bioenergy sources is the preparation of liquid and gaseous bioenergy sources as fuels for vehicles. As this is only an intermediate step, this application area is often little different from combined heat and power. The use of fuels for vehicles is in many cases based, like power generation, on combined heat and power engines.

2.13.1 Heat generation

Heat can be generated using all bioenergy sources in solid, liquid or gaseous state. Although the amount of heat generated depends only on the calorific value of the particular fuel that is used, the basic conditions required for complete combustion with low emissions differ a great deal depending on the aggregate state.

2.13.1.1 COMBUSTION OF SOLID BIOENERGY SOURCES

Solid organic fuels have the property of not being flammable themselves under ambient conditions. In order for a solid bioenergy source to burn, a highly complex chain of thermochemical conversion processes needs to take place:

1. heating
2. drying
3. pyrolytic decomposition
4. gasification of the water-free fuel
5. gasification of the solid carbon
6. oxidation of the combustible gases.

The technical requirements for a full conversion of solid fuels in this process chain are as follows:

- The oxidant air must be supplied in excess (more than stoichiometric).
- The process control must bring about a sufficiently good mixing of the fuel gases and the supplied combustion air.
- The fuel gas/air mixture generated in the process requires a sufficiently long dwell time in the reaction zone.
- The whole process requires a sufficiently high combustion temperature.

Modern solid fuel boilers are designed to create these suitable technical conditions. In essence they achieve this by a spatial separation of the air supply to the firebed (primary air inlet) and the air supply to the gas combustion zone (secondary air inlet). This guarantees even combustion of the fuels and a low level of emissions.

The individual phases in a fire using solid fuels are as follows:

- Phase 1: Warming of the fuel (less than 100°C). When fed into combustion systems, solid fuels are generally at the ambient temperature of the storage facility – that is, at a temperature ranging between 10°C and 25°C. Before further reactions can begin, the solid fuel needs to go through a phase of warming.
- Phase 2: Drying of the fuel (between 100°C and 150°C). Above 100°C there begins the main vaporization of water adhering to and trapped inside the fuel. This escapes from the fuel as water vapour.
- Phase 3: Pyrolytic decomposition of the wood components (between 150°C and 230°C). Pyrolytic decomposition begins at temperatures around 150°C. In this process the long-chain components of the solid fuels are broken down into short-chain compounds. Products that arise as a result of this are liquid tar compounds and gases such as carbon monoxide (CO) and gaseous hydrocarbons (C_mH_n). The pyrolytic decomposition of wood does not require any oxygen.

Phases 1–3 are endothermic (heat-absorbing) reactions. They take place in any fire automatically, and serve to prepare the fuel for oxidation. Once the flash point is reached, which is at around 230°C, the field of influence of exothermic (heat-giving) reactions begins, with the input of oxygen. Wood can be externally ignited at around 300°C, and from 400°C undergoes spontaneous combustion.

- Phase 4: Gasification of the water-free fuel (between 230°C and 500°C). The thermal decomposition of the water-free fuel under the influence of oxygen starts at the flash point of around 230°C. The gasification takes place mainly in the firebed of a solid fuel fire. Here the oxygen supplied as primary air gives off sufficient heat in its reaction with the gaseous pyrolysis products to be able to affect the solid and liquid pyrolysis products such as carbon and tar.
- Phase 5: Gasification of the solid carbon (from 500°C to 700°C). In this phase, under the influence of carbon dioxide (CO_2), available water vapour and oxygen (O_2), combustible carbon monoxide is generated. The gasification of solid carbon is exothermic and gives off light and heat rays that take the form of a visible flame.

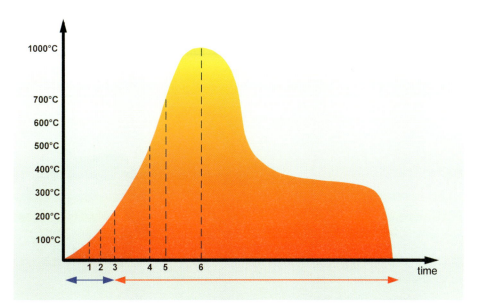

*Figure 2.46.
Temperature graph of a wood fire
Graph: Dobelmann/www.sesolutions.de*

■ Phase 6: Oxidation of the combustible gases (from 700°C to around 1400°C). The oxidation of all combustible gases resulting from the preceding process stages represents the end of the combustion reaction for the solid fuels. Under the influence of secondary air, the clean and complete combustion of the gas mixture is brought about.

For a wood fire, the conversion process can be visualised as in Figure 2.46. Wood fires that are ignited with a charge of fuels have the heat curve shown here. The combustion of solid materials is based on a balance between endothermic (heat-absorbing) reactions, shown here with a blue arrow, and exothermic (heat-giving) reactions, shown here with a red arrow. It can be seen here that firing with solid fuels, in contrast to constant gas or oil flames, means longer heat-up times and a fluctuating heat output.

Figure 2.47 shows a log fire. The different colours of the flames in a wood fire are the result of various combustion processes. Yellow flames occur with the post-combustion of carbon particles such as soot. Blue flames occur when the wood is pyrolysed to carbon monoxide. Both volatilization and the following combustion phase are strongly dependent on the available reaction surface.

*Figure 2.47.
Flames of a wood fire
Photo: creativ
collection/www.sesolutions.de*

When a large piece of wood is burned, this happens in a continuous process in which the thermochemical changes move from the outside to the inside of the material. Figure 2.48 shows a simplified cross-section through a burning piece of wood. Here we can see how the various process stages of drying, gasification and combustion move through the wood. Consideration of these processes reveals that the available reaction surface plays an important role in the speed of these processes.

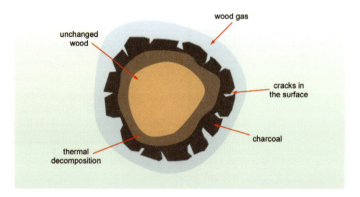

Figure 2.48.
Cross-section through a burning log
Graphic:
Dobelmann/www.sesolutions.de

Because reduction of the size of the fuels increases the specific surface area for the reactions to take place, swifter conversion of the fuel is possible. If a heater is to give out its warmth in allotted bursts, the fire needs to have short starting times and short afterglow phases. By increasing the surface area it is possible to create the right conditions for this (Figure 2.49).

Figure 2.49.
Surface reaction area/volume-masses
Graphic:
Dobelmann/www.sesolutions.de

Splitting and reducing the size of firewood creates the ideal conditions for running a low-emissions combustion system, in which the start-up and burning-out phases of the fuel are minimized. This makes the heat output of combustion systems easier to dispense as required, and heat storage systems can be designed more accurately for the heat absorption.

As the combustion proceeds through the various process stages, pollutants are given off from the solid biofuels. However, good process control can eliminate these before they can escape to the environment. Pollutants from the combustion of solid bioenergy sources can be divided into two classes:

- pollutants resulting from incomplete combustion
- pollutants resulting from complete combustion.

Harmful substances resulting from incomplete combustion are carbon monoxide (CO), soot (C), hydrocarbons and tar compounds (C_mH_n), and unburned particles. The creation of these pollutants can be avoided if the combustion meets the following criteria:

- minimum temperature > 800°C
- sufficient oxygen, air surplus > 1.5 Lambda value (combustion air over supply)
- dwell time of the gases in the combustion zone > 0.5 s.

Pollutants resulting from complete combustion comprise mainly nitrogen oxides (NO_x) and residual carbon monoxide (CO). Whereas nitrogen oxides are generated both from the combustion air and from the nitrogen content of the fuels, carbon monoxide is an indicator of the combustion quality. The carbon monoxide content in the exhaust gases from combustion systems is determined mainly by the air surplus figure (λ) and by the design of the combustion system.

Modern combustion systems with separate routeing of the primary and secondary air and a sufficiently large combustion zone create the right conditions for low-emission combustion. In this regard, hand-fed or automatically fed boilers are used in applications for pure heat production from solid bioenergy products.

For good overall efficiency, modern combustion systems have to create the optimum process conditions for all phases. This applies particularly to the primary and secondary supplies of combustion air, which represent the limiting factor for clean combustion and high combustion efficiency.

In modern boilers, both air feeds are designed with blowers that are often electronically controlled, or fitted with adjustable air flaps. This enables the output of the boiler to be varied with the primary air feed. With a secondary air feed regulated in the same way, it is possible to ensure optimum gas combustion at all times. As a consequence, the pollutant emissions of these boilers are minimized in running operation and also particularly in the critical warm-up phase.

In modern houses, heat is required either evenly, such as for room heating, or rapidly, such as to heat water for the shower. These technical requirements pose big problems for unregulated wood heating systems. In practice, this problem is solved either by sufficiently dimensioned storage tanks or by a constant feeding of new fuel with an automatic feed system.

Well-dimensioned storage tanks (Figure 2.50) can absorb the entire heat production of one charge of fuel and store it in their contents, which they can then deliver to the house's heating network when required.

Figure 2.50.
Modern storage tank
Photo + graphic: Viessmann
Werke/www.viessmann.com

The way a storage tank works in conjunction with a wood boiler can be explained with the aid of the graph shown in Figure 2.51. Modern storage systems signal to the boiler via a thermostat when they need heat. This allows the combustion intervals of the boiler to be coordinated better and the number of ignition sequences in the boiler to be reduced.

A storage system enables wood heating systems to be flexible and react to short-term requirements for heat, and also extends the combustion intervals. This helps to reduce stresses on the wood boiler and minimizes the number of partial load firings. Storage systems, especially with hand-fed central heating boilers or the integration of a solar installation, form an integral part of the heating system.

Another possibility for achieving a constant heat output with the use of wood is to use an automatic feed unit. Systems with automatic feed, such as woodchip or pellet boilers, always request new fuel before their output drops (Figure 2.52).

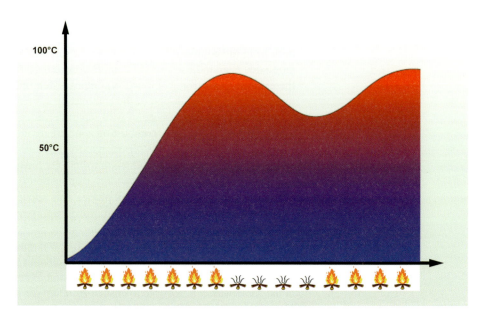

*Figure 2.51.
How a storage system works
Graph: Dobelmann/www.sesolutions.de*

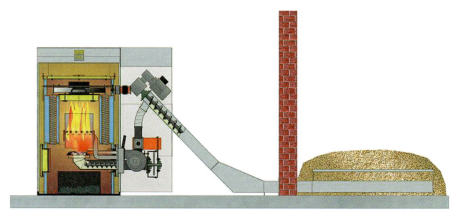

*Figure 2.52.
Automatically fed boiler
Graphic: Oekofen/www.okeofen.at*

Manually fed boilers for solid wood or pellets are available with a power range from 1 kW up to 100 kW thermal output. Automatically fed boilers are grouped into a 10–50 kW range that uses mainly wood pellets, and an output range above 50 kW in which the use of woodchip boilers is predominant. However, the output range of automatic heating supply stations running on solid biomass goes significantly beyond this. For example, biomass furnaces with combined local heat can provide heat output of up to several megawatts.

2.13.1.2 COMBUSTION OF LIQUID BIOENERGY SOURCES

Liquid bioenergy sources such as vegetable oil or biodiesel can also be used for pure heat production.

In its viscosity and burning behaviour, biodiesel is the same as conventional heating oil. Therefore oil-fired boilers, provided all plastic parts and metals in contact with the fuel are resistant to biodiesel, can also be used for biodiesel without any conversion (Figure 2.53).

Natural vegetable oil, by contrast, has a higher viscosity than heating oil, and hence different burner geometries are required for its use. It cannot be used in conventional oil burners.

Some manufacturers offer special rapeseed oil burners that, for example, use a centrifuge method to ensure an ignitable distribution of the viscous oil. Even if mixtures of up to 20% vegetable oil in heating oil do not significantly affect the resulting viscosity, any such process cannot generally be recommended. Even with a mixture proportion of 5%, carbonization and deposits on the nozzles and orifice plates have been observed in practice.

*Figure 2.53.
Modern oil-fired boiler
Graphic: Viessmann
Werke/www.viessmann.com*

The market for stationary utilization of vegetable oil has so far produced only a very few applications, with the result that use in practice cannot draw on comprehensive operating experience.

2.13.10.3 COMBUSTION OF GASEOUS BIOENERGY SOURCES

Biogenic gases, if of suitable quality, can be used in conventional gas boilers. Low-temperature and condensing boilers (Figure 2.54) can be used here.

*Figure 2.54.
Modern gas condensing boiler
Graphic + photo: Ritter GmbH & Co.
KG/www.paradigma.de*

Biogas has a lower rate of flame propagation than natural gas, and special burner jets therefore need to be fitted in boilers using biogas. For small heating outputs up to 30 kW, atmospheric burners are generally used. Larger quantities of gas, on the other hand, can only be burned using blower burners.

The fatigue life and maintenance intensity of heaters running on biogas are strongly dependent on the gas composition as found following its preparation. In condensing boilers, in particular, a high residual content of hydrogen sulphide (H_2S) can lead to serious corrosion damage.

Biogas and other biogenic gases such as wood gas are rarely used purely to generate heat. It is often economically more interesting to turn the energy content into electricity. This is generally implemented in a combined heat and power system.

2.13.2 Combined heat and power

Electricity from the utilization of bioenergy sources is generated primarily in combined heat and power (CHP) generators. In terms of the technology for use in combined heat and power, in addition to combustion engines, turbines, Stirling and steam engines and to a small extent fuel cells also are used.

The background to combined heat and power lies in economic concerns and in energy efficiency. The generation of electricity from fuels for the most part takes place via CHP generators, which extract mechanical energy from the thermal energy. This can be converted into electrical power in generators. In practice the maximum attainable conversion efficiency for electricity is around 40%. The rest of the energy continues to exist as heat.

In a CHP system, electricity and heat are generated at the same time. In contrast to a condensing power station, in which the heat produced in CHP generators is dissipated via a heat exchanger, for combined heat and power in CHP plants or cogeneration plants electrical and thermal energy are utilized immediately.

If the aim is to design the utilization of bioenergy sources to be economically optimum, the heat arising during power generation needs to be utilized. A decentralized CHP system can achieve an overall process efficiency of 90% of the primary energy input. This is a big increase in the utilization efficiency compared with the mere 36% efficiency in centralized electrical energy generation in condensing power stations (Figure 2.55).

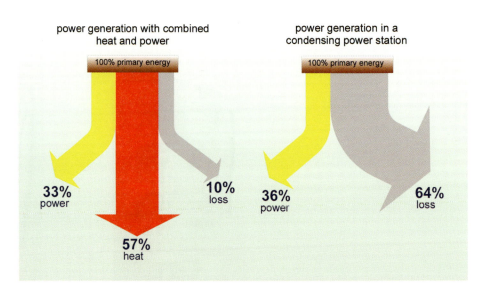

Figure 2.55.
Efficiency of power generation
Graphic:
Dobelmann/www.sesolutions.de

Power generation with power–heat coupling achieves better utilization of the primary energy input. In the case of fossil energy sources the main importance of this is in terms of climate protection. Bioenergy sources benefit from the increased efficiency of this method.

2.13.3 Dimensioning of the combined heat and power system

The dimensioning of the electrical and thermal output of cogeneration plants is a decisive factor for the efficiency of the project as a whole. Generators that run in grid-parallel operation – that is, connected directly to the electricity grid – can be designed to be heat controlled. In this case the system is regulated in such a way that the generated heat is utilized, and if, at the same time, electricity is generated surplus to internal requirements, it is output to the electricity grid.

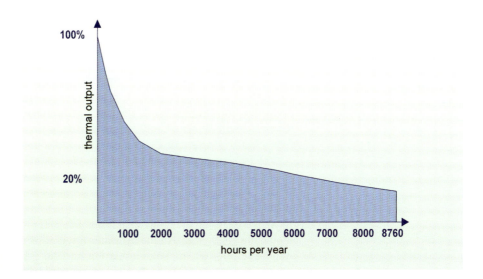

Figure 2.56.
Annual load duration curve for heat requirements
Graph: Dobelmann/www.sesolutions.de

The basis for the dimensioning of this kind of CHP system is an annual load duration curve. The energy requirements of the heating network at a location are calculated over the year and represented in the form of a graph (Figure 2.56). Here the thermal output used in a building is recorded and plotted in descending order in a graph. CHP systems, also called cogeneration units, have the optimum dimensioning when they have the greatest possible annual running times in the provision of heat.

For economic reasons, CHP systems are dimensioned to cover the basic load. This allows running times of over 5000 hours per year to be reached. The investment costs are thereby distributed over a larger energy volume, and the electricity supplied from the CHP system is cheaper per unit.

CHP systems that are dimensioned according to economic considerations cover some 10–35% of the maximum thermal or electrical energy that a site uses. As a rule, this achieves an annual running time of over 5000 hours. In terms of the total power, the CHP system then covers around 50–80% of annual power and heat requirements.

If a CHP system is to be installed in combination with a boiler, in practice a thermal dimensioning of the CHP system to 20% of the boiler output makes sense. This guarantees an annual running time for the unit of over 5000 hours and therefore economical operation.

Table 2.7.
Dimensioning cogeneration systems
Data: Glizie GmbH/www.glizie.de

Annual heat requirements (kWh/yr)	Heat requirements in August (kWh)	Annual electricity requirements (kWh)	Optimum cogeneration output (kW$_e$)
150,000	4500	45,000	7
250,000	7500	75,000	11
400,000	**12,000**	**120,000**	**18**
600,000	18,000	180,000	27
800,000	24,000	240,000	36
1,000,000	30,000	300,000	45
1,200,000	36,000	360,000	54
1,400,000	42,000	420,000	63
1,600,000	48,000	480,000	72
1,800,000	54,000	540,000	81
2,000,000	60,000	600,000	90
2,200,000	66,000	660,000	99
2,400,000	72,000	720,000	108
2,600,000	78,000	780,000	117
2,800,000	84,000	840,000	126
3,000,000	90,000	900,000	135
4,000,000	120,000	1,200,000	180
5,000,000	150,000	1,500,000	225
6,000,000	180,000	1,800,000	270
8,000,000	240,000	2,400,000	360
10,000,000	300,000	3,000,000	450

The values in Table 2.7 are a guide for dimensioning a CHP system for inhabited buildings such as elderly care homes, hotels, hospitals and blocks of flats. The dimensioning is worked out as follows.

A cogeneration system has optimum output when the annual heat requirements, the heat requirements in the month of August, and the annual electricity requirements are reached or exceeded in the CHP electrical output line. Note here that the respective lowest line value of the items determines the power output of the system. In the example shown in bold for an apartment building, this results in an optimum CHP electrical output of 18 kW.

Precise dimensioning is important in the planning of cogeneration plants. If the selected unit is too large, it may well be uneconomic, as its annual running times are too low. Conversely, if the selected unit is too small, then the full potential economic and ecological benefit is not achieved.

When planning the implementation of cogeneration plants, a detailed assessment should always be made of the hydraulic integration into the heating grid, the electrical systems technology, the noise insulation, and the exhaust gas routeing and fuel supply.

As an example of the hydraulic integration of CHP units into thermal building networks, a coupling with low-temperature boilers and with condensing boilers was selected.

Example 1: CHP system and low-temperature boiler
When using a CHP system together with conventional boilers or low-temperature boilers or wood boilers, the integration of the system into the return of the heating circuit to the boiler brings technical advantages. In this way, a CHP system can be integrated even into existing heating circuits without requiring special control technology. This solution is engineered by tapping the return pipe to the boiler and feeding heating water heated in the CHP unit into the main return to the boiler. Running the CHP unit therefore raises the return temperature to the boiler, thus reducing the boiler operating times. See Figure 2.57.

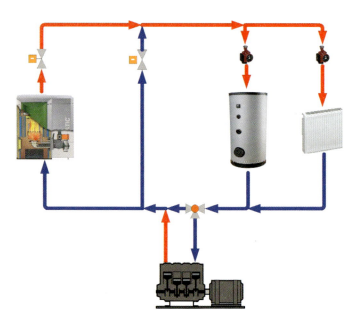

Figure 2.57.
Cogeneration system in heating circuit with wood-fired boiler
Graphic:
Dobelmann/www.sesolutions.de

Example 2: CHP with condensing boiler and storage tank
If a CHP unit is used in conjunction with a condensing boiler, it is disadvantageous to integrate the CHP unit into the return, as this limits the heating energy utilization of the boiler. The CHP unit is therefore integrated into the heating system parallel to the condensing boiler. A storage tank or hydraulic switch is used for hydraulic equalization. This method as far as possible avoids the pumps interfering with each other. With this method of operation, the boiler pump should have a temperature-dependent control. When the boiler is operating, this allows the avoidance of any overflow of hot water via the storage tank into the boiler. See Figure 2.58.

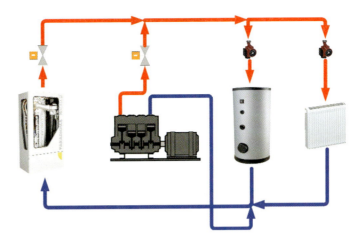

*Figure 2.58.
Cogeneration system in heating circuit
with condensing boiler
Graphic:
Dobelmann/www.sesolutions.de*

The CHP unit's return connection is located on the load side so as not to force the CHP unit to switch off while the boiler is operating even if there is an overflow via the storage tank.

As well as the cogeneration system and the boiler, storage tanks are used in many applications. This is justified by their use to cover short-term peak heat loads. With their heat storage volume they avoid switching on the peak load boiler, and increase the continuous running time of the CHP unit. This avoids cyclical operation at short time intervals and increases the service life of the CHP unit.

In practice, the dimensioning of the heat-exchange capacity of the heat exchanger to the warm water storage tank should be matched to the thermal output of the CHP unit and not to the total thermal output of the heating system.

2.13.3.1 COMBUSTION ENGINES

A technological variation on the use of liquid and gaseous bioenergy sources in CHP generation is combustion engines. These units exist in two different technological groupings according to the type of combustion: internal and external combustion.

Engines with internal combustion include gas spark ignition engines, gas-diesel engines and pilot injection diesel engines. An example of an engine with external combustion is the Stirling engine. Table 2.8 summarises the technical features of the various engine systems. The choice of engine to use depends mainly on the project. The engines are chosen for their power and specifications in conjunction with the available fuel.

*Table 2.8.
Technical data for combustion engines
Graphic:
Dobelmann/www.sesolutions.de*

	Gas spark ignition engine	Gas-diesel engine	Pilot injection diesel engine	Stirling engine
Combustion	Internal	Internal	Internal	External
Efficiency (%)	22–27	> 35	28–35	< 30
Service life	Low	High	Medium	Experimental
Maintenance requirements	High	Low	High	Experimental
Investment costs	Low	High	Medium	Experimental
Performance class (kW)	>5	> 150	30–150	Experimental

Small projects with low and possibly sporadic heat requirements tend to be equipped with gas spark ignition engines or pilot injection diesel engines. High-compression gas-diesel engines, which are expensive in terms of investment costs,
tend to be used in large projects.

2.13.3.1.1 GAS SPARK IGNITION ENGINE

In their construction, gas spark ignition engines are similar to the petrol engines used in cars. The carburettor used in petrol engines is replaced by a gas mixer. This generates an ignitable mixture of the fuel gas and the engine's intake air, which is then ignited by an ignition spark from the spark plug in the cylinder. See Figure 2.59.

*Figure 2.59.
Gas spark injection cogeneration system with 300 kW_e output
Photo: Dobelmann/www.sesolutions.de*

2.13.3.1.2 GAS-DIESEL ENGINE

Gas-diesel engines can either be designed as normal diesel engines, which are then fitted with a spark ignition, or their compression is so high that spark ignition is not necessary. The power class of these engines is generally in excess of 150 kW electrical power. As a result of the robust construction of these large industrial engines, service lifetimes of over 80,000 operating hours are normal. See Figure 2.60.

*Figure 2.60.
Fitting a piston in a large engine
Photo: MAN BW/manbw.de*

2.13.3.1.3 PILOT INJECTION DIESEL ENGINE

Pilot injection diesel engines are converted standard diesel engines that are able to burn biogas mixed with the intake air (Figure 2.61).

As this air/gas mixture does not auto-ignite at the compression pressures created in the diesel engines, external ignition has to be provided, as for gas spark ignition engines. Pilot injection diesel engines use the existing injection nozzles and introduce diesel and heating oil into the cylinder along with the compressed gas/air mixture. This jet of fuel ignites as a result of the compression and consequently ignites the mixture, and the combustion process takes place. The amount of firing oil required to run the engine in this way is around 7–10% of the total engine output attained. The service life of this engine type is around 30,000–40,000 running hours.

*Figure 2.61.
Pilot injection diesel cogeneration
system with 100 kW$_e$ power
Photo: Dobelmann/www.sesolutions.de*

2.13.3.1.4 STIRLING ENGINE

Stirling engines are an example of engines that use external combustion. These heat- and power-generating machines utilize the temperature difference between two points and convert this energy difference into mechanical energy. See Figure 2.62.

*Figure 2.62.
Stirling cogeneration system
with 10 kW$_e$ output
Photo: Dobelmann/www.sesolutions.de*

Stirling engines are not yet widely found in industrial applications. Unfortunately, the advantage of external combustion in practice also often causes problems with ensuring a constant heat transfer to the cylinder. For this reason, Stirling engines in the bioenergy field have still not advanced past the status of research applications.

2.13.3.1.5 STEAM PISTON ENGINE

Steam piston engines (Figure 2.63) are the modern version of the steam engine invented by James Watt in 1769.

*Figure 2.63.
Steam piston engine with
1.5 MW$_e$ output
Photo: Spilling Energie/www.spilling.de*

Engines driven by steam are fed from steam boilers. They require steam pressures between 6 and 60 bar. If the engineering intention is for the engine to be an intermediate element in a production circuit, then back-pressures of up to 25 bar can be tolerated. Steam engines are able to handle steam flow rates of up to 40 t/h.

They are designed to be oil-free, and can be supplied in various sizes. The performance classes are between 25 kW and 1500 kW per individual engine. Steam engines have nominal engine speeds of between 750 and 1500 rev/min, and can be supplied for other technical applications apart from power generation with variable speeds.

If it is necessary to increase the power of the engines, they can also be arranged in a cascade circuit, and it is possible to work with multi-stage expansion models. The power yield of the engines can also be optimized with a fill control for full and partial load ranges.

Steam engines are robust and well-proven units for solid fuels. They are easy to operate and, because of their slow piston speeds and low wear, they offer high availability. Steam engines have low requirements for feed water quality, and can be used efficiently everywhere that combined heat and power is desired with a proportion of power of approximately 10–15%.

2.13.3.2 TURBINES

Turbines can basically be divided into four different types:

- steam turbines
- gas turbines
- organic Rankine cycle (ORC) turbines
- hot air turbines.

The sections below provide a technical explanation of these turbine concepts.

2.13.3.2.1 STEAM TURBINES

Steam turbines (Figure 2.64) are the traditional way of using fuels of various origin for generating electrical energy.

In the steam power process, process water is fed via a water feed pump into a steam boiler, consisting of a vaporiser and a superheater. Here it changes from a liquid state into a gaseous state. The volume of steam that results is then expanded via a turbine that is coupled to a generator. The cooled and expanded steam is recooled in a condenser or cooling tower and returned to a liquid state.

Figure 2.64.
Assembling a steam turbine
Photo: Siemens AG/www.siemens.de

The structural expense involved in the units required for steam generation means that, for economic reasons, a lower power limit for this process is 300 kW of electrical power.

2.13.3.2.2 Gas turbines

Unlike the steam turbines described in the previous section, which operate in a closed, externally heated circuit, gas turbines (Figure 2.65) are fired directly. Here the fuels are combusted with oxygen in a combustion chamber forming part of the turbine and expelled across the turbine blades. The resulting rotary motion is converted into electrical energy by a generator.

Figure 2.65.
Assembling a gas turbine
Photo: Siemens AG/www.siemens.de

Gas turbines are produced in sizes up to several megawatts. However, micro gas turbines (Figure 2.66) can also provide lower electrical power ranges starting at 30 kW.

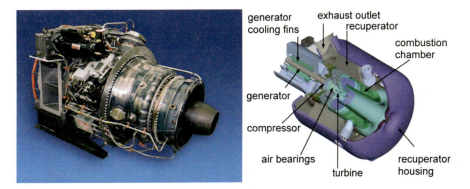

Figure 2.66.
Micro gas turbines
Photo: MTU AG/www.mtu.de. Graphic:
Capstone Inc./www.capstone.com

Typically, these heat- and power-generating machines suffer large efficiency losses in partial load operation, so that it is necessary to ensure an even incoming flow to the turbine with a constant volume of gas. Overall, the electrical efficiency of small gas turbines below an electrical power level of 200 kW, at an average of 25–29%, is below that of conventional combustion engine cogeneration systems.

2.13.3.2.3 Organic Rankine cycle turbines

In organic Rankine cycle (ORC) systems, turbines are used that, instead of water, use low-boiling-point organic solvents as the working medium. Their power range is from 50 kW up to 2.5 MW.

ORC turbines can be used with lower temperature differences than is the case with the other turbine systems discussed in this section.

As a typical secondary process, application areas are the use of geothermal and solar thermal energy, and the utilization of waste heat from biomass heating stations.

2.13.3.2.4 Hot air turbines

In some cases the direct combustion of biogenic gases in open gas turbines causes technical problems. In these cases, as well as steam or ORC turbines, hot air turbines with external combustion can also be used.

These turbines differ from the internal combustion gas turbines in that they have external firing to heat the working medium, which is air. Here the air, routed in a circuit, is brought up to the turbine entry temperature with nearly isobaric heating. In the turbine, the air then undergoes an irreversible expansion back to atmospheric pressure.

As for all turbines, the net power results from the difference between the turbine power and the compressor power. Hot air turbines are typical primary processes that require a high temperature level.

2.13.3.3 FUEL CELLS

Fuels cells are a type of electrochemical energy converter. They can convert hydrogen-rich gases into water with oxygen directly from the air or in pure form, and extract electrical energy and heat directly from this process. This form of electrochemical conversion was recommended as long ago as 1897 by Wilhelm Ostwald at the founding meeting of the Bunsen Society for the Resource-Conserving Generation of Electrical Energy from Fuels.

Figure 2.67.
Fuel cell in industrial use
Photo: MTU AG/www.mtu.de

Four types of fuel cell have technology suitable for the utilization of biogenic gases:

- polymer electrolyte membrane fuel cell (PEMFC)
- phosphoric acid fuel cell (PAFC)
- molten carbonate fuel cell (MCFC)
- solid oxide fuel cell (SOFC).

The two additional existing types, alkaline and direct methanol fuel cells, cannot be operated directly with the main component of biogas, methane (CH_4). Alkaline fuel cells (AFC) can be operated only with pure hydrogen (H_2). Direct methanol fuel cells (DMFC) are also ruled out for the use of biogas, as they can run trouble-free only if they have pure methanol as their fuel. Hydrogen and methanol can be extracted from solid biomass. Even if the processes used to do this, such as thermochemical gasification and pyrolysis, are for the most part still in the research stage, in the long term it will be possible to use these fuel cells with biogenic fuels.

2.13.4 Processing into a product

Apart from direct energy utilization of liquid and gaseous bioenergy sources for heat generation or combined heat and power, the preparation and sale of processed fuels is an interesting alternative. The quality requirements for preparation and the resulting product quality are particularly high for this type of application.

2.13.4.1 PROCESSING INTO VEHICLE FUEL

The various options for processing liquid bioenergy sources into fuels are explained in depth in the section on liquid bioenergy.

If it is desired to use biogenic gases as fuels, they need to be processed and compressed to 200 bar in order to ensure a sufficient operating range for the fully fuelled vehicle. As well as measures to inhibit corrosion, such as removing hydrogen sulphide and ammonia, it is also essential to filter and dry the gas that is used. A further measure is to separate out the carbon dioxide in order to increase the calorific value of the fuel generated.

Various different processes can be used to prepare biogas for this purpose. One option is to use molecular sieves to separate the carbon dioxide, in combination with active carbon filters to eliminate the hydrogen sulphide. Another option is to use absorption pressure washers, powered with water, to simultaneously eliminate both gas components.

The minimum quality requirements for the use of biogas as a fuel in vehicles are listed in Table 2.9.

Table 2.9.
Quality requirements for biogas fuel

Name	Unit	Biogas raw gas	Biogas fuel
Methane (CH_4)	Vol.%	50–75	>95
Carbon dioxide (CO_2)	Vol.%	25–50	3–4
Water vapour (H_2O)	g/m^3	10–50	0.032
Nitrogen (N_2)	Vol.%	0–5	–
Oxygen (O_2)	Vol.%	0–2	<1.0
Hydrogen (H_2)	Vol.%	0–1	<0.5
Ammonia (NH_3)	Vol.%	0–1	–
Hydrogen sulphide (H_2S)	Vppm	0–6000	<15
Solid particles	μm	<100	<5

2.13.4.2 FEEDING INTO THE NATURAL GAS GRID

For feeding biogas into natural gas or town gas grids, the same preparation mechanisms need to be used as described in the previous paragraph for the preparation of fuels. Because the pressure existing in biogas facilities is insufficient, or can be subject to large fluctuations due to the generation process, before feeding into the public gas grid it is necessary to increase pressure to the relevant pressure ranges:

- low-pressure lines: up to 50 mbar
- medium-pressure lines: 50 mbar to 1 bar
- high-pressure lines: over 1 bar up to 80 bar.

Because biogas has to go through all the stages of treatment on site and be adapted to the prevailing natural gas quality in the grid with a separation of carbon dioxide, direct grid feeds are so far something of a rarity.

3 Anaerobic digestion

3.1 Introduction

Anaerobic digestion (AD) of manure is a technique that has been applied for several decades. Whereas the first AD installations experienced various technical difficulties, nowadays it is considered to be a technically proven and commercially attractive way to produce renewable energy. The advantages of biogas plants are manifold:

- economically attractive investment
- easily operated and safe installation (no technical background required)
- production of renewable electricity and heat, resulting in a reduction of CO_2 emissions
- reduction of methane emissions from manure storage
- improvement of the fertilizing qualities of manure.

Thousands of biogas installations have been built around the world in the last few years, ranging from household-sized digesters (mainly in developing countries) to large-scale centralized digesters integrated in a manure treatment plant. In this chapter on AD we shall focus on farm-scale digesters. Generally speaking, AD at a farm scale can be interesting for farms, users of manure or other parties that have a minimum of 1000 t of manure per year at their disposal.

3.1.1 Who should read this chapter?
This chapter is meant for farmers, project developers, investors and any other person or party who is interested in the realization of an anaerobic digester on a farm-scale level.

It will provide an insight into the technical aspects of an anaerobic digester, the economical feasibility of a specific application, the legal aspects related to AD, and the effort required for operation and maintenance.

3.1.2 What information will be provided?
In section 3.2 the functioning of an anaerobic digester and its possible configurations (focusing on combined heat and power generation) are discussed. The biological process is explained, together with the corresponding process conditions. Biogas production from manure and other possible organic matter will be considered, and also the quality of the digested manure.

In section 3.3 the planning of an AD project is discussed. This is done step by step, starting with the creation and definition of the project, then considering system choice, layout and sizing, and finally concluding with an installation that is ready for realization. The various parties that should be consulted at each stage of the planning process, such as suppliers of equipment and electricity companies, are discussed. Aspects concerning the actual commissioning and start-up are presented in section 3.4. The activities that are required for the operation and maintenance of an anaerobic digester are discussed in section 3.5. In section 3.6 a tool that can be used to estimate the investment costs and assess the economic feasibility of the proposed installation is presented, including a detailed example calculation for a specific case.

All legal aspects related to a farm-scale anaerobic digester are discussed separately in Chapter 8.

3.2 System description and components

In general, the principle of all anaerobic digesters is the same. Manure and other possible biomass (co-substrates) are inserted into a large, sealed, airless container. In this oxygen-free environment, bacteria will produce biogas. In most digesters the contents will be heated to accelerate the process.

The produced biogas can be used to generate heat or electricity, or both. This last option, combined heat and power (CHP) mode, is the most common. The electricity that is generated by the gas engine can be either supplied to the electricity grid or used for own consumption. The heat is used for the digester, and the surplus heat can be used, for example, for heating stables or for residential heating

AD can be applied at a range of scales, depending on the amount of biomass available. Systems can range from small, farm-scale digesters to large centralized anaerobic digesters (CADs) supplied with feedstock from several sources. In this chapter the focus is on farm-scale digesters using CHP.

3.2.1 System description

3.2.1.1 SYSTEM OVERVIEW

Figure 3.1 shows an overview of a typical anaerobic digester. The various components are described below.

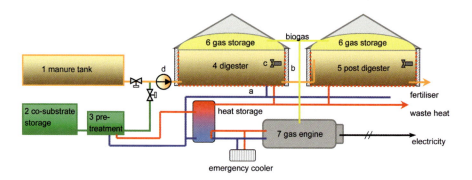

Figure 3.1.
Schematic overview of a typical AD system: (a) digester heating; (b) digester insulation; (c) stirring device; (d) substrate pump
Graphic: Ecofys bv / www.ecofys.com

3.2.1.1.1 MANURE STORAGE

The most common manure storage systems are cellars, silos and manure bags. Manure contains bacteria that will produce methane as soon as it is excreted (by means of cold digestion). Methane production in storage will lower the biogas yield of the digester. In the case of open cellar storage, methane emissions are also undesirable for animal well-being. Therefore it is best to transport the manure from the storage to the digester as soon as possible. This is generally done by means of a pump.

3.2.1.1.2 STORAGE CO-SUBSTRATES

The addition of other biomass with a higher energy density than manure can substantially increase the biogas yield. The additional biomass is called co-substrate. The difference in fluidity of the co-substrates, as compared with the manure, will generally warrant separate storage.

3.2.1.1.3 PRE-TREATMENT

Basically there are three different methods for pre-treatment, depending on the type of co-substrate and the size characteristics as delivered to the AD plant: mechanical treatment, preheating and thermal treatment.

Some co-substrates require a size reduction, such as chopping or grinding, to prevent the occurrence of pieces of co-substrate that are too large for the pumps and mixers of the installation. Size reduction also increases the surface area for the bacteria, which will accelerate biogas production. Other types of co-substrate, such as

*Figure 3.2.
Overall view of a farm-scale AD plant
Photo: Smack AG /
www.schmack-biogas.com*

fats, may require preheating in order to improve the flow characteristics. Certain co-substrates need thermal treatment in order to fulfil sanitation requirements (see section 3.2.2.6.3).

3.2.1.1.4 Digester

In the digester the substrates are heated and the fermentation process takes place. The two end products of this process are biogas and digested substrate. The contents are stirred periodically. This is necessary for the following reasons:

- to mix new substrate with the old substrate to improve the penetration of bacteria with the fresh substrate
- to realize an even temperature in the digester
- to prevent and disturb the build-up of sedimentary layers
- to improve the metabolism of the bacteria by removing the gas bubbles and replacing them with fresh feedstock.

3.2.1.1.5 Post-digestion storage

The digested substrates (the digestate) are commonly stored in a post-digestion storage tank (see section 3.2.4.11). This serves as a storage tank for the digested substrate, as not all of it can be used directly once digested. Moreover, additional biogas is produced therein.

3.2.1.1.6 Biogas storage

The biogas that is produced in the digester will have to be stored until it is used. It can be stored either in the digester or in an external gas storage unit. This will depend on the chosen construction (see section 3.2.4.8).

3.2.1.1.7 The gas engine

The gas engine, functioning as a CHP unit, will use the biogas to generate electricity and heat. The generated electricity can either be used for own purposes or be supplied to the grid (or a combination of both). The produced heat will partly be used for heating the digester. The remaining heat can be used for the heating of buildings or stables, or for other purposes such as greenhouses or industrial processes.

These various components are discussed separately in section 3.2.4.

Biogas can also be processed for use as a transportation fuel, for supply to a natural gas grid, or for generating heat only. In developing countries biogas produced by unheated digesters is used as a cooking fuel. In this chapter, however, we shall focus on heated farm-scale digesters with a CHP unit.

3.2.1.2 SIZE

We can distinguish between the following sizes of AD system:

- Small scale. These are simple digesters with a capacity (5–100 m^3) for small quantities of substrates (100–1000 t). In Europe digesters of this size are often not profitable because of the high investment costs of the technical components in relation to the relatively low yield; they can mainly be found in Asia. These digesters have no insulation, heating or stirring facilities.
- Farm scale. Farm-scale digesters have a capacity of 100–800 m^3 and can process 1000–15,000 t of substrates a year. A large part of these substrates will generally originate from a single farm. In most cases the electricity produced will be provided to the grid. Residual heat can be used as a substitute for other heat-producing energy sources. See Figure 3.3.

Figure 3.3.
A farm-scale AD plant
Photo: PlanET GmbH /
www.planet-biogas.com

- Industrial scale. An industrial-scale digester has a capacity of over 15,000 t of substrates. Because of the scale, this type of AD application often offers economically attractive opportunities for further treatment of the digestate – for example to produce high-quality liquid fertilizers. There are also industrial-scale biogas installations, which digest wet organic wastes, such as waste water from their own processes, organic waste from food processing, or the separated organic fraction of municipal solid waste. See Figure 3.4.

Figure 3.4.
An industrial-scale AD plant
Photo: ABR Agrar Bio-Recycling GmbH/
www.abr-w.de

3.2.2 Biogas from manure and co-substrates

3.2.2.1 THE BIOLOGICAL PROCESS

During the process of AD, bacteria decompose the organic matter in order to produce the energy necessary for their metabolism. Methane is produced as a by-product of this metabolism. Figure 3.5 shows the main theoretical stages and the intermediary products of the AD process. In practice, these stages coexist within the process, and each stage is characterised by the main activity of a specific group of bacteria.

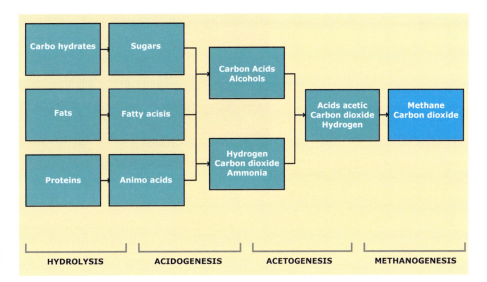

Figure 3.5.
Schematic diagram of the main stages of the AD process[1]
Graphic: Ecofys bv / www.ecofys.com

3.2.2.2 PROCESS CONDITIONS

The bacteria that produce the methane (see Figure 3.6) can only live in a specific environment. The following aspects are important:

- Anaerobic conditions. The bacteria are active only when no oxygen is present.
- Moist conditions. A moisture content in the substrate of at least 50% is necessary.
- Temperature. There are basically three temperature ranges in which specific strains of bacteria are most active. These are psychrophilic (<30°C), mesophilic (30–40°C) and thermophilic (40–55°C) strains. The thermophilic strains have the highest activity. Most farm-scale digesters operate in the mesophilic range. This process is less sensitive to changes and therefore more easily controlled than the thermophilic process.
- Retention time. The retention time is the time for which the substrates are inside the digester. The required retention time for optimal biogas production depends on the temperature (see Figure 3.7). For biogas production that is near the theoretical maximum, the psychrophilic temperature range requires a retention time of 40–100 days, the mesophilic range 25–40 days, and the thermophilic range 15–25 days.
- pH. The pH value in the anaerobic digester should be around 7.5. Special attention should be given to this aspect in the case of co-digestion of acid substrates (such as some wastes from the food processing industry).
- Organic load. The bacteria need a minimum organic load (organic dry matter per m^3 reactor per day) as 'food' to survive, but they can also be 'overfed'. The organic load should be between the extremes of 0.5 kg and 5 kg of organic matter per m^3 of digester tank per day ($OM/m^3/day$). A healthy situation would be between 1 and 3 $kgOM/m^3/day$. In order to avoid too high an organic load at the feed-in point of the digester, feeding of fresh substrate into the digester should be done at least daily.
- Auxiliary substances. The bacteria need soluble nitrogen compounds, minerals and trace elements as auxiliary substances for their metabolism. When manure is used as a major component of the substrate, these substances are present in sufficient quantities.

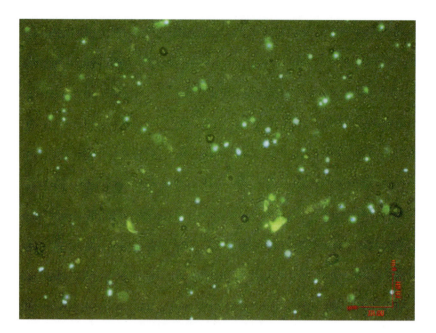

*Figure 3.6
Microscopic view of digestion bacteria
Photo: Smack AG /
www.schmack-biogas.com*

- Restraining substances. Some substances that may occur in manure, such as disinfectants, antibiotics and organic acids, impede the bacteria in their activity or even kill them. High concentrations (for example as a result of cleaning all the stables with disinfectant at once, or treatment of the whole livestock with antibiotics) should be avoided.
- Particle size. The particles in the substrate should not be too large; otherwise the bacteria do not have sufficient surface to attach to. It may be necessary to reduce the size of specific co-substrates such as grass or sugar cane.
- Mixing of the substrate. The gas that is produced by the bacteria will only surface automatically if there is less than 5% dry matter in the substrate. In all other cases, mixing is necessary to avoid pressure build-up.
- Consistent conditions. Rapidly changing process conditions should be avoided. Feeding of fresh substrate into the digester should be done gradually. This also applies to a change in substrate composition, for example as a result of a change in forage.
- Nitrogen content. The presence of nitrogen in the substrate is necessary because it is an essential element for the bacteria's metabolism, and it helps to maintain the pH (when converted to ammonia it neutralises acids). However, too much nitrogen in the substrate can lead to excessive ammonia formation, resulting in toxic effects. A healthy carbon to nitrogen (C:N) ratio is between 20:1 and 40:1, although more extreme values can still result in efficient digestion.

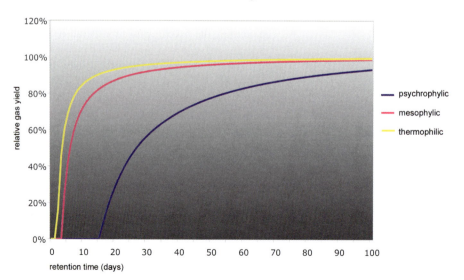

*Figure 3.7.
The relation between temperature and retention time
Graphic: Ecofys bv / www.ecofys.com*

3.2.2.3 BIOGAS COMPOSITION

The produced biogas consists of methane (CH_4) and carbon dioxide (CO_2), together with minor quantities of nitrogen, hydrogen, ammonia and hydrogen sulphide (see Table 3.1).

Table 3.1. Composition of biogas

Component	Volume percentage
Methane (CH_4)	50–80%
Carbon dioxide (CO_2),	50–20%
Nitrogen (N_2)	<1%
Hydrogen (H_2)	<1%
Ammonia (NH_3)	<1%
Hydrogen sulphide ($H2_S$)	<1%

The methane content of the biogas varies between 50% and 80%. The higher the methane content of the biogas, the more energy it contains. Biogas from substrates with high carbohydrate content, such as cattle manure and corn, has a relatively low methane content.

3.2.2.4 BIOGAS YIELD

When all the process parameters are within the required range, the biogas production will be near the theoretical maximum. The biogas yield is determined by the characteristics of the substrate. The following properties are of importance:

- dry matter (DM): percentage of dry matter in the substrate
- organic matter (OM): the organic fraction (%) of the dry matter
- organic dry matter (ODM): the organic part of the substrate (= DM × OM)
- maximum specific biogas production (in m^3/t ODM).

The total biogas production can be calculated using the following formula:

Biogas production = amount of substrate (t) × DM (%) × OM (% of DM) × maximum biogas production (m^3/tODM)

Example
1000 t of pig manure has a dry matter (DM) content of 8%, of which 80% is organic matter (OM). The maximum biogas yield is 450 m^3/tODM. The biogas production from digestion of this manure will be: 1000 t × 8% DM × 80% OM × 450 m^3/tODM = 28,800 m^3 biogas.

3.2.2.5 SUBSTRATES

Substrates are the materials that will be digested. On a farm, the basic substrate is manure, but other 'wet' organic material, such as corn or grass, can also be added.

3.2.2.5.1 MANURE

The composition of manure varies by type of animal and by farm. This has a large effect on the biogas yield. Table 3.2 gives an overview of the variation in composition and biogas yield of manure from various animals. The dry matter fraction can vary as a result of water use on a farm, and the biogas yield can vary for different types of feed.

Table 3.2. Characteristics of manure from different animals

Feedstock	Dry matter, DM (%)	Organic matter (% of DM)	Biogas yield (m^3/t ODM)	Biogas yield (m^3/t wet)	Average biogas yield (m^3/t wet)
Cow manure	7–15	65–85	200–400	9–51	25
Pig manure	3–13	65–85	350–550	7–61	27
Chicken manure	10–20	70–80	350–550	24–88	51

*Figure 3.8.
Manure from cows, pigs and chicken
are the basic substrates for AD on farms
Photo: PlanET GmbH /
www.planet-biogas.com*

Manure from dairy cows has a much lower biogas yield per kg organic dry matter than manure from pigs. This is due mainly due to the cows' paunch flora, which stimulates digestion of the manure to start before excretion. However, the relatively high dry matter content of their manure will compensate for this.

Dairy cows produce around 27 t of manure per animal per year. However, depending on the pasturing policy on a specific farm, much of this amount is produced when they are pasturing. Sows produce around 5.5 t of manure per animal per year, and hogs around 1.2 t. Manure from sows has a lower dry matter content than manure from hogs, and therefore a lower biogas yield per tonne.

3.2.2.5.2 Co-substrates

The addition of co-substrates (see Figures 3.9 and 3.10) to manure (co-digestion) is an economically attractive way to increase biogas production. Co-substrates generally have a substantially higher biogas yield per (wet) tonne than manure, and they can be acquired from various sources. On most farms there will be some leftover silage or other agricultural waste. It is also possible to grow crops especially for the purpose of AD (so-called energy crops). In many cases co-substrates originate from external sources – for example, waste originating from food processing industry.

*Figure 3.9.
Pictures of three different co-substrates
that can be used for anaerobic digestion
with manure
Photo: PlanET GmbH /
www.planet-biogas.com*

When co-substrates are added, the required process conditions, such as organic load or pH (as set out in section 3.2.2.2), must be considered carefully. Also, most farm-scale biogas systems are not capable of dealing with dry matter contents (of the mixture of manure and co-substrate) of over 15%. As most cow and pig slurry has a significantly lower dry matter content, it is possible to add co-substrates with higher dry matter contents. Table 3.3 lists the composition and yield of various co-substrates. This table should be used as an indication only, as the composition of co-substrates may fluctuate in practice.

*Figure 3.10.
Picture of substrate in digester
Photo: PlanET GmbH /
www.planet-biogas.com*

*Table 3.3.
Characteristics of some co-substrates*

Feedstock	Dry matter, DM (%)	Organic matter (% of DM)	Biogas yield (m^3/t ODM)	Biogas yield (m^3/t wet)	Average biogas yield (m^3/t wet)
Vegetable waste	10–20	65–85	400–700	25–120	75
Mangold	10–20	80–95	800–1200	65–230	145
Corn silage	15–40	75–95	500–900	55–340	200
Grass silage	30–50	80–90	500–700	120–315	220
Fat and flotation slurry	8–50	70–90	600–1300	30–585	310

Figure 3.11 shows the average biogas yield per tonne of wet material, and the variation for some possible substrates.

When using co-substrates, it is important to consider the aspects mentioned in the next section, and also specific legal requirements such as additional permits that may be required (see Chapter 8).

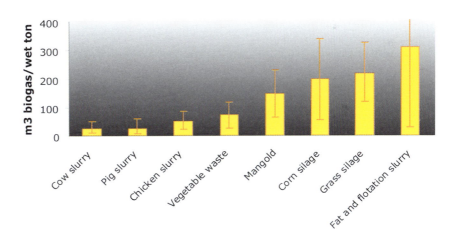

*Figure 3.11.
Average biogas yield per tonne of
wet material, and the variation
for some substrates.
Graphic: Ecofys bv / www.ecofys.com*

3.2.2.6 PROPERTIES AND QUALITY OF (CO-)DIGESTED MANURE

3.2.2.6.1 Digested manure

Digested manure has several advantages over untreated manure:

- The fraction of nitrogen (N) that is directly absorbed by the plants has increased. This is a result of the conversion of easily degradable organic compounds. With the right use of the digested manure (for example, manuring at the beginning of the growing season to avoid wash-out of N), it is possible to save on chemical fertilizer use. Savings of 10–20% have been achieved in practice.

- The organic compounds that degrade very slowly (the humus-like compounds, also called lignin) are not degraded in the AD process. Therefore the function as a soil improver is maintained.
- Digested manure has fewer odours.
- The digested manure is more homogeneous.
- The amount of pathogens and seeds in the manure is reduced. This is discussed further in section 3.2.2.6.3.

3.2.2.6.2 Co-substrates

When using co-substrates there are several things to look out for. Apart from meeting the process conditions (as set out in section 3.2.2.2), the following aspects should be considered.

- Chemical aspects. Co-substrates may contain heavy metals (such as zinc and copper) or other inorganic contaminants and persistent organic pollutants (POPs). When digestate (the digested mixture of manure and co-substrates) is to be used on agricultural land, it is advisable to check whether the concentrations of these contaminants in the digestate meet the required national and/or regional standards (see Chapter 8). The co-substrates may also contain N, P and K, sometimes in significantly higher concentrations as in manure. This must be accounted for when the digestate is used as a fertilizer.
- Physical impurities. It is possible that physical impurities may be present in the co-substrate. These could include plastic and rubber, metal, glass and ceramic, sand and stones, cellulose materials (wood, paper, etc.), and other impurities. These impurities can affect the operational stability of the installation, or even damage the plant components. Most of these impurities are also undesirable if the digestate is used on the agricultural land. It is advisable to ensure that the co-substrate has as few physical impurities as possible and meets the national and/or regional standards for this (see Chapter 9).
- Pathogens and seeds. Co-substrates may contain various pathogens and seeds, depending on their source of origin. Co-substrates originating from external sources may also pose an additional risk for spreading diseases (such as BSE, foot and mouth, or potato rot) or for causing weeds, especially when the digestate is used on agricultural land. This risk varies for the various types of co-substrate. This is discussed further in the next section.

3.2.2.6.3 Pathogens and seeds

As a result of the heating of the substrate in the digester, a large proportion of the pathogens and seeds are killed. The higher the process temperature, the higher is the degree of reduction. In Table 3.4 this effect is shown for some pathogenic bacteria that are present in manure.

As a result, digested manure contains fewer pathogens and seeds than undigested manure. However, co-substrates originating from external sources may contain additional pathogens and seeds.

A classification of the various types of co-substrates with respect to potential risk of pathogens is[1]:

Table 3.4. Comparison between the decimation times (T90) of some pathogenic bacteria in digested and untreated slurry[2] The decimation time is the time in which the bacteria are reduced by 90%.

Bacteria	Digested slurry		Untreated slurry	
	53°C (hours)	35°C (days)	18–21°C (weeks)	6–15°C (weeks)
Salmonella typhimurium	0.7	2.4	2.0	5.9
Salmonella dublin	0.6	2.1	–	–
Escherichia coli	0.4	1.8	2.0	8.8
Staphylococcus aureus	0.5	0.9	0.9	7.1
Mycobacterium paratuberculosis	0.7	6.0	–	–
Coliform bacteria	–	3.1	2.1	9.3
Group of D–Streptococci	–	7.1	5.7	21.4
Streptococcus faecalis	1.0	2.0	–	–

- sludge from vegetable production
- sludge etc. from fish farming
- sludge etc. from animal production
- source-separated waste (e.g. from households)
- sewage sludge.

The digestate of sludge from vegetable production or from fish farming is unlikely to cause any risk of pathogens. The other categories do pose an additional risk of pathogens, and therefore require sanitation. In most cases, sanitation is executed by heating the slurry to 70°C for 1 h (pasteurization). The sanitation process reduces the pathogens to a satisfactory level. Sanitation takes place in a small, separate tank that is heated. Sanitation of the fresh feedstock (before the digestion process) is called pre-sanitation. Post-sanitation is done after the digestion process.

If there is uncertainty about the types of seeds in the co-substrate, it is also advisable to apply sanitation. As an example, verge grass presents a relatively low risk of spreading of animal diseases caused by pathogens, but it may sometimes contain high levels of seeds. The sanitation process reduces the seeds to a satisfactory level.

3.2.3 Various AD systems

The basic idea of each anaerobic digester is the same. Substrates are added to a sealed container in which biogas is produced by means of a digestion process. The biogas is stored in a tank in order to achieve a constant supply to the CHP unit. There is a large variety of different AD systems. Each system has its own advantages and disadvantages. The most common systems will be discussed in the following section.

3.2.3.1 ANAEROBIC DIGESTION PROCESSES

AD systems can be considered under the following three categories (see Figure 3.12) according to their process management[3]:

- continuous processes
- semi-continuous processes
- discontinuous processes (batch systems).

3.2.3.1.1 Continuous processes

A biogas plant running in continuous mode comprises the main digester and a separate post-digestion tank for the digestate. Existing manure storage tanks can be used as post-digestion tanks. When substrates are added to the digester a similar amount of digestate will flow to the post-digestion tank via an overflow pipe. Therefore the level in the digester remains constant. The digestate that enters the post-digestion tank may contain some substrates that will continue to digest in the storage. Commonly the post-digestion tank is also air sealed, so that biogas produced here can be used as well in order to increase the overall gas production. Another option to improve the overall system efficiency is to use the methane-rich air from the post-digestion tank as combustion air for the gas engine.

This continuous process is most suitable for farmers who have to store their manure for long periods. The digester can be constructed on a relatively small scale, because it will contain the manure only for the duration of the digestion process (retention time). The storage function of the digester is less important, as the digestate is stored mainly in the post-digestion storage.

3.2.3.1.2 Batch process

This is another common option for operating an anaerobic digester. The digester is periodically completely filled with manure and co-substrate; it is then sealed and the digestion process starts. The biogas production rate increases over time until it reaches a maximum. When the production rate has subsequently dropped below a certain level, 90–95% of the digestate is moved to the storage tank. The rest remains behind in the digester to start the digestion process for the next batch of fresh substrate. In order to achieve a constant biogas supply it is necessary to have several digesters in

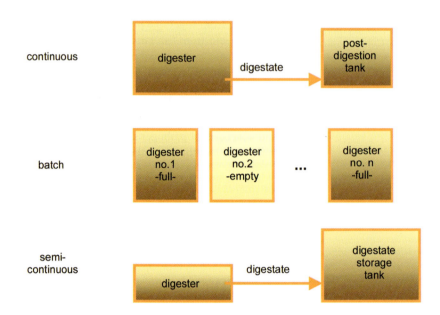

Figure 3.12.
Schematic overview of AD processes
Graphic: Ecofys bv / www.ecofys.com

different stages of the AD process, operating in parallel. Commonly systems of multiple digester tanks are more suitable for industrial large-scale AD plants; two-tank batch digestion is also found at the farm scale.

3.2.3.1.3 Semi-continuous process

This type of process combines the advantages of the batch and the continuous processes. It allows use of the digester for both storage and digestion of substrates. The organic material is continuously added to gradually fill the digester tank. The digested manure will be kept in the tank as long as storage is required. Once the digester is full it turns from being a batch-operated digester into a continuously operated digester. Thus any additional substrate that is added from that stage on leads to an outflow of digestate to the digestate storage tank. As for continuous processes, existing manure storage tanks are commonly used as storage tanks. These tanks are generally large enough that no additional post-digestion tank is needed. The major disadvantage of this process originates from the fact that a part of the digestate will not be completely digested, and hence the biogas yield is lower than for other process options. Owing to the shorter retention time the sanitation effect is slightly lower than for the batch or the continuous process.

3.2.3.2 PRINCIPLES OF DIGESTION

There are two basic types of digester: horizontal and upright. Semi-continuous systems commonly use horizontal digesters. For continuous digesters both types can be applied.

3.2.3.2.1 Horizontal digester

Horizontal digesters (Figure 3.13) are normally relatively small. They consist of a large steel tank with a stirring system. The standard volume of a horizontal digester is between 50 and 150 m^3. Normally these tanks are transported to the site in one piece. Therefore their size is limited by the maximum sizes allowed for road transport.

When the substrates are inserted into the digester the heating arms heat it. The heating arms are mounted on the axle of the mixer, and therefore rotate in the substrate. This type of digester always requires external gas storage.

The substrate enters one side and the digestate leaves the other side. A quantity of substrates will move through the digester at an even pace. The advantage of this type of digester is that the substrates are not mixed horizontally, but only vertically. Therefore the sanitation effect and the average biogas yield are higher. The retention time can be shorter, which increases the capacity of the digester.

A horizontal digester can handle a dry material percentage of 15–20%.

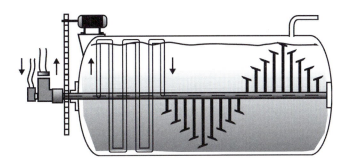

Figure 3.13.
Horizontal digester
Graphic: Ecofys bv / www.ecofys.com

3.2.3.2.2 Upright digester

An upright digester (Figure 3.14) is cylindrical in shape, and normally has a volume between 300 and 1500 m^3. The content is heated either by an external heater, which heats the ingoing substrates, or by hot water via tubes alongside the wall. The walls are insulated to reduce the loss of heat. There are various stirring systems possible. In most cases the biogas will be stored above the digestate, under a flexible membrane roof.

A horizontal digester can also be fitted with a solid covering, using external gas storage to store the biogas.

In most cases an upright digester will be less expensive than a horizontal digester because of the use of cheaper materials such as concrete, and because of the less complicated construction. The stirring of the substrates can be done using several different types of stirring device; they are discussed in detail in section 3.2.4.5.

An upright digester can handle a dry material percentage of up to 10–15%.

Figure 3.14.
Upright digester
Photo: Smack AG /
www.schmack-biogas.com

3.2.3.2.3 Important selection criteria for digester type

- The size and type of the current manure storage. Possibly the existing manure storage can be used as a digester tank or post-digestion storage.
- The dry material percentage of the substrate. A horizontal digester can handle a maximum dry material percentage of 15–20%, whereas an upright standard digester can handle a maximum dry material percentage of 10–15%. It is common to use an upright digester when the dry material percentage is below 10%.
- The desired retention time of the digester. If the digester also has to serve as temporary manure storage it is advisable to use a semi-continuous digester.
- The investment costs of the digester. A horizontal digester is relatively expensive. The existing storage systems might be used as post-digestion storage. In the case of an upright digester it might also be possible to use the existing manure storage as digester tank.

3.2.3.3 COMMONLY USED AD SYSTEM LAYOUTS

There is considerable variety in the designs of AD system layouts. Each has its advantages and disadvantages. Important reasons for selecting a particular layout are:

- the availability of substrates
- the available investment resources
- the available infrastructure (for example a silo to be retrofitted into a digester)
- the available space
- the required sanitation
- the climate (a cold climate requires better insulation)
- the required (or preferred) storage time of the digestate
- preference for a supplier.

In Figure 3.15 an overview is presented of several possible system layouts. There is a distinction made between digesters with internal biogas storage and those with external storage.

System variants a and b in Figure 3.15 are semi-continuously operated digesters, in the most simple and economic construction. Owing to the good cost/performance ratio, in many cases new digesters will be constructed as shown in variant c. The relatively expensive digester can be sized as small as possible.

Sometimes both the digester tank and the post-digestion storage are used as biogas storage (Figure 3.15e). Horizontal digesters will often be designed as in Figure 3.15d and 3.15f. For design option c and d a range of post-digestion storage systems can be used (see section 3.2.4.11). System variants g and h display common batch digestion systems.

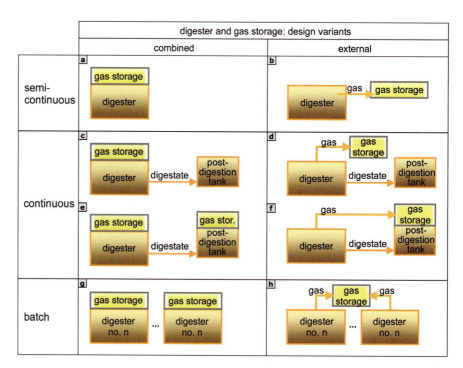

Figure 3.15.
Schematic overview of typical AD system layouts
Graphic: Ecofys bv / www.ecofys.com

3.2.4 System components

3.2.4.1 DIGESTER TANKS

In this chapter the focus is on upright digesters. The floor and walls of upright digesters are usually made of steel-reinforced concrete, but steel can also be used. The walls of the digester need to be covered with insulation material to prevent the loss of heat.

The digester must be air sealed. The top cover of the container depends on the type of storage applied. For integrated gas storage the cover may consist of a flexible, synthetic gasproof foil. The disadvantage of this construction is the low insulating effect and thus the relatively high heat loss. The use of a solid roof provided with insulation can prevent this. In most cases, however, this will require external gas storage.

3.2.4.2 DIGESTER HEATING AND INSULATION

The process parameters of anaerobic digestion demand temperatures that are commonly above the ambient temperature. Mesophilic anaerobic digestion operates between 25°C and 35°C, and thermophilic processes above 40°C. Therefore it is necessary to heat the substrate up to the required temperature. In general, heat generated in the cogeneration engine is used for this. It is transferred from the engine to the digester by standard heat pipes. See Figure 3.16.

*Figure 3.16.
Digester heating
Photo: PlanET GmbH /
www.planet-biogas.com*

There are three basic types of substrate heating, depending on the type of digester. In horizontal digesters the heating is integrated in the stirring device. Heating of vertical concrete digesters is commonly installed as wall heating, with the heat piping attached to the inside wall of the digester. Here stainless steel piping offers several advantages compared with PVC piping, including excellent thermal conductivity and a lower tendency to fouling. Unlike concrete digesters, metal digesters can be equipped with outside wall heating. Floor heating can also used, but this is questionable, as the sediment layer on the bottom of the digester has reduced thermal conductivity characteristics, and hence acts almost like insulation.

Biogas plants with sanitation tanks, in which part of the substrate is heated up to 70°C, can often dispense with digester heating, as they use a counterflow heat exchanger to heat the main stream of the substrate exchanging heat with the sanitised substrate. In this case the insulation of the digester should be a bit thicker than usual.

In order to reduce heat losses from the substrate via the digester walls it is necessary to insulate the digester (Figure 3.17). Common insulation materials are mineral wool, mineral fibre mesh, expanded or extruded polystyrene, and polyurethane foam. Alternatively organic materials made from cotton, wool, cork or similar materials can be used to insulate the digester. Polyurethane is used mainly for the side walls of a digester, with a thickness of about 6 cm, whereas polystyrene is commonly applied to the bottom of a digester, with a thickness of about 8 cm. Mineral wool is used for both the bottom and side walls of a digester, with a thickness of about 10 cm.

The selection of the appropriate insulation material depends on the digester size and the specific price of the insulation material. Each insulation material has specific thermal conductivity characteristics that influence the thickness that needs to be used (see Table 3.5). In designing the digester insulation the goal should be to find the economic optimum balancing between the cost of the insulation material and the savings due to reduced heat losses. In addition a top layer will be required to protect the insulation against the influences of dirt and weather.

*Figure 3.17.
Digester insulation
Photo: Krieg & Fischer Ingenieure
GmbH / www.kriegfischer.de*

*Table 3.5.
Characteristics of various
insulation materials[3]*

Insulation material	Density (kg/m^3)	Heat conductivity (W/m K)
Polyurethane		0.030
Expanded polystyrene	20–45	0.040
Extruded polystyrene	30–80	0.035
Mineral wool	30–50	0.043
Cork	100–120	0.050
Sheep wool mesh	10–20	0.035
Cotton mesh	20	0.040

3.2.4.3 PIPING FOR SUBSTRATE TRANSPORT

See Figure 3.18. There are two types of piping: pressurized piping (for transportation purposes) and non-pressurized piping.

Piping that is pressurized to transport substrate, by means of a pump, must have a diameter of at least 100 mm to prevent blockages. At longer distances a diameter of at least 150 mm is required[3]. In order to prevent sediments on the bottom of the piping a minimum transport speed of about 1 m/s should be considered. Non-pressurized piping is subject to the influence of gravity. It requires a diameter of at least 200 mm.

Commonly the piping used for AD plants is made of steel. However, other materials such as plastics are also used.

*Figure 3.18.
Substrate piping
Photo: Smack AG /
www.schmack-biogas.com*

Any piping that is exposed to frost must be protected by means of insulation, in order to prevent ice formation and thus blocking of the piping.

A check valve should be installed to prevent the contents flowing back from the digester to the manure storage.

3.2.4.4 PUMP

See Figure 3.19. A pump in an AD system may have two functions: it can either serve to overcome a difference in height, or it is can be used to drive a hydraulic stirring system (see section 3.2.4.5).

The two main types of pump are centrifugal pumps and displacement pumps: the latter can be classified as eccentric spiral pumps, vane pumps, or bellow pumps.

Figure 3.19.
Different types of pump
Photo: PlanET GmbH /
www.planet-biogas.com

In order to facilitate the operation of pumps, piping should be installed with a declining slope of 1–2%, so that the pump will be emptied automatically during standstill. This prevents the build-up of sediment in the pump. However, this is not possible with displacement pumps since they block the outlet flow when not running.

3.2.4.5 STIRRING DEVICES

The operation of stirring devices fulfils various purposes in a digester:

- balancing of the temperature of the substrate
- mixing of old substrate with new, so that active bacteria are present everywhere in the substrate
- prevention of the formation of agglomeration and layers.

Figure 3.20.
Screw propeller
Photo: Krieg & Fischer Ingenieure
GmbH / www.kriegfischer.de

Figure 3.21.
Hydraulic stirring
Photo: PlanET GmbH /
www.planet-biogas.com

Stirring devices in AD plants can be classified into mechanical and hydraulic stirring devices. Common mechanical stirring devices are screw propellers (Figure 3.20).

A screw propeller system consists of an electric motor with a load capacity of 2.5–25 kW, which drives a screw propeller. This system is operated by hand and is suitable for digesters up to a volume of 1000 m^3. The propeller creates a flow in any desired direction. To prevent and disturb the creation of sedimentary layers a height adjustment is necessary. This is normally designed as a height-adjustable propeller or a swivel arm propeller.

Alternatively the substrate can be stirred hydraulically (Figure 3.21) by pumping it out of the digester at one point and feeding it back at another point. The feedstock is normally sucked into a pipe in the upper part of the digester and ejected into the lower part. The input and output of the pipe must be placed in such a way that the complete content is mixed. In most cases the substrate feed-in pump can be used for this purpose. To make this possible the transport tube is diverted by means of a valve and an additional tube.

This system is suitable only when fluid co-substrates are used. It is less suitable for co-substrates that tend to form sedimentary layers.

The advantage of a hydraulic system is that there are no moving parts in the digester. The pump is located outside the digester and is thus easily accessible for maintenance.

3.2.4.6 SUBSTRATE STORAGE

See Figure 3.22. The most commonly used manure storage systems are cellars, silos, basins and manure bags. When a digester is used it is advisable to store the manure for as short a time as possible before it is fed into the digester because the digestion process starts during storage, leading to reduced biogas yields in the digester.

Figure 3.22.
Substrate storage systems
Photo: PlanET GmbH /
www.planet-biogas.com

The storage of the co-substrate will depend largely on its physical and chemical properties. For example corn can be stored as silage, but fats are likely to require a storage tank (possibly with a heating device to assure their liquid state).

3.2.4.7 CO-SUBSTRATE FEEDING SYSTEM
See Figure 3.23. Many solid co-substrates will require a size reduction treatment before insertion into the digester. The particle size of the co-substrate must be small enough that it can be added to and mixed with the manure. This requires a feeding system that will chop or grind the co-substrates. The substrates can be inserted directly into the digester tank using a dry matter input system, e.g. a funnel-shaped tank. Fluid substrates can also be inserted directly into the digester tank, possibly using the storage system.

Figure 3.23.
Co-substrate feeding system
Photo: A) PlanET GmbH / www.planet-biogas.com B) Ecofys bv / www.ecofys.com

Alternatively, a pre-mixing well can be used to mix the manure and co-substrates before pumping into the digester.

In order to ensure a good control of the amount of co-substrates supplied, a dose and weighing system is necessary.

3.2.4.8 BIOGAS STORAGE
Biogas will normally be stored at atmospheric conditions. This requires a larger volume than conventional gas storage cylinders. The actual volume is determined by the biogas yield and the consumption pattern. If an engine is running constantly it will require smaller gas storage than an engine that operates only at peak electricity demands.

Gas storage tanks can be distinguished by their operating pressure; common differences exist between low-pressure storage, intermediate and high-pressure tanks. Low-pressure storage operates just above ambient pressure, and is commonly made of flexible foils. The operating pressure is set by a throttle valve placed along the gas piping that leads to the CHP engine, and depends also on the weight of the foil. Intermediate (5–20 bar) and high (200–300 bar) pressure tanks are designed as steel pressure vessels and gas bottles. Low-pressure biogas storage is most commonly used for farm-scale AD plants and operates at underpressures between 0.05 and 0.5 mbar. It is described further below.

As discussed in section 3.2.3, biogas storage can be internal (on top of the substrate or digestate) or external. If the biogas is stored internally, a flexible foil is installed above the feedstock. The flexible foil, a membrane about 1–2 mm thick, will expand when biogas is created. A foil will not be necessary when the required gas storage is small. If the biogas is stored externally, a gas bag can be used. These bags store biogas at low pressure without stressing the bag material, which ensures a long lifetime.

Figure 3.24.
Biogas storage: some variants
Photo: Krieg & Fischer Ingenieure
GmbH / www.kriegfischer.de

Gas bags have several advantages compared with other external storage types. They can be manufactured on site in any size, even up to 2000 m³, at reasonably low costs, and the foil is resistant against corrosion. However, such gas bags must be protected from destruction and weather influences.

3.2.4.9 THE BIOGAS ENGINE

The chemical energy stored in the biogas is commonly transferred into heat and power by using conventional gas engines (combined heat and power, CHP) (Figure 3.25). In this chapter only the most frequently used gas engine, a piston engine that drives an electrical generator, will be discussed. Other biogas conversion equipment for generating electricity includes the Stirling engine, fuel cells, and the gas turbine. However, these options are not yet commercially proven (for this application). Recently there have been some investigations into the use of fuels cells for farm-scale digesters.

Normally a piston engine releases heat to the atmosphere via the engine cooling water and the exhaust system. In a CHP configuration this heat is recovered by means of heat exchangers. Part of it is used to heat the digester. The remaining heat can be used to meet an external heat demand. CHP engines can make use of up to 90% of the fuel's energy content, converting it to about 30% electric energy and 60% heat.

Figure 3.25.
CHP engines
Photo: Smack AG /
www.schmack-biogas.com

The generated electricity can either be used on site or supplied to the grid. There are two different options for electricity generation:

- Constant nominal production. For the optimum lifespan of the engine it is best to keep the motor running constantly. In this configuration the CHP unit will be as small as possible, thereby minimising investment costs. This, however, leaves no possibility of enlarging the electricity output capacity at times of peak electricity demand.
- Demand driven. The engine will operate primarily when the electricity demand is the highest (peak hours). During this period electricity generally has the highest value, either by using it on site or by supplying it to the grid.

Engines for biogas are based on standard four-stroke engine types that are produced in volume. Two-stroke engines are not suitable, because of their high wear levels. In most cases a suitable engine will be bought from a company that specializes in adapting engines to run on biogas.

A specific type of piston engine is the dual fuel engine. This starts on diesel. Once the engine is running, biogas is added, and the amount of diesel is reduced to 10–20% of the engine's fuel usage. This is the minimum amount required to ignite the mixture and lubricate the engine. The biogas is mixed and then sucked into the engine. The advantage of a diesel engine is that it can run on a mixture with a relatively low CH_4 content. Furthermore, it can function as an emergency power unit. The disadvantage is that the engine will emit fossil (non-renewable) carbon monoxide (up to 10 times more than gas engines fuelled by biogas only).

In order to start up a digester, heat is required to heat the digester content. A dual fuel engine can run on diesel during start-up, and produce hot water until the biogas production starts.

Electricity generated with a dual fuel engine is not considered renewable energy in some countries (see Chapter 9).

3.2.4.10 DESULPHURIZATION

Biogas may contain about 1% hydrogen sulphide (H_2S). This has a corrosive effect on metals, and will therefore damage the engine and the piping. It is therefore important to remove the H_2S. This can be done simply by adding some air (2–6 vol.%[4]) into the upper part of the digester, close to where the biogas outlet to the biogas storage is located. See Figure 3.26. Oxidizing bacteria will convert the hydrogen sulphide into sulphur, after which it will drop into the digestate as elementary sulphur. When the amount of air is dosed correctly, the amount of H_2S in the biogas can be reduced by 95%. However, if too much air is added the elementary sulphide will convert into sulphuric acid. Furthermore, the combination of air and biogas can be very explosive. Therefore it is very important to limit the quantity of air that is added.

Figure 3.26.
Desulphurization
Photo: PlanET GmbH /
www.planet-biogas.com

An H_2S measuring instrument can monitor the performance of the desulphurization. An aquarium air-pump can serve as the simplest pump to add air in an easy and controllable way. This simple, reliable and low-cost method is commonly used in farm-scale digesters.

3.2.4.11 POST-DIGESTION TANK/DIGESTATE STORAGE TANK

After the substrate has been sufficiently fermented it is transferred to the post-digestion tank (Figure 3.27) to be stored until the digested substrate can be used, for example as fertilizer. Increasingly these tanks are covered, in order to avoid nitrogen losses and to collect the additional biogas that forms during storage of the digestate. Commonly it is not permitted to distribute manure on the fields during the colder months of the year. Accordingly this storage tank should be designed to be large enough to accommodate the amount of digested substrate that is produced during a period of about 6 months. Oversizing the post-digester tank offers slight advantages regarding later extension of the biogas plant. It is common for the existing manure storage tanks to be used as post-digestion tanks.

Figure 3.27.
Post-digestion tank
Photo: PlanET GmbH /
www.planet-biogas.com

3.2.4.12 MEASUREMENT AND CONTROL EQUIPMENT

There are several measurement devices that enable the operator of a biogas plant to run the system efficiently and thus ensure the plant's economic success (Figure 3.28). They also facilitate the daily control of the performance of the various plant components and the detection of malfunctioning or misbehaviour of the system. The most important measuring devices are as follows:

- Temperature sensors. These are usually attached to the wall of the digester to measure the temperature of the substrate. Also, in order to determine the consumption of the produced heat, temperatures need to be measured at the advance and backflow of the heating network; if they are used in combination with a flowmeter the generated heat and process heat consumption can be calculated.
- Substrate level indicators. A level indicator helps the operator to assess the performance of the digester and thus the amount of biogas produced. It is important to know the daily amounts of added substrate and thus the substrate flow.
- Electricity meters. Two meters are required: one to measure the internal consumption of the system and one to measure the electricity delivered to the grid.
- Gas meters. At least two gas meters should be installed, to measure the gas production and the gas consumption; the measured gas flows are indicators for the performance of the biogas plant. Moreover, they help in the safe operation of the system.

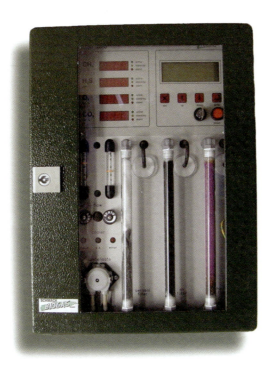

*Figure 3.28.
Measurement equipment
Photo: Smack AG /
www.schmack-biogas.com*

The various measuring devices can also be connected to a computer in order to automate data acquisition and operation.

There are additional parameters that it is useful to measure regularly.

- pH. The pH value of the substrate and the digestate is important in ensuring the appropriate living environment for a good performance of the bacteria.
- Composition of the biogas. In particular the methane and hydrogen sulphide (H_2S) contents are indicators of the performance of the digester, and are necessary control parameters to ensure desulphurization of the biogas and thus avoid corrosion.
- Dry matter content. The dry matter content of the substrate is important in estimating the biogas yield and the processing rate.
- Ammonia. The ammonia concentration also influences the biogas production rate of the bacteria; at higher concentrations this rate decreases. In the digested material the ammonia concentration gives an indication of its potential as a fertilizer.
- Fatty acids. The concentration of short-chained fatty acids gives an indication of the performance of the digestion process, and allows the operator to react to changes in the digestion environment, as the lower the short-chained fatty acid concentration is, the more compounds can be in the substrate that are toxic to the bacteria.

In general these parameters are not measured continuously in farm-scale biogas plants. However, the pH of the substrate and the digested material should be checked daily in order to detect changes in the sensitive environment of the bacteria. This measurement can be carried out by using litmus paper or simple electronic pH-meters. Ideally the biogas composition should be measured online by a gas chromatograph, as the detection of H_2S is important in ensuring a long lifetime for the cogeneration engine. In practice the determination of the biogas composition is usually reduced to measuring the CO_2 content with Brigon CO_2-indicators, and H_2S with special tubes whose content reacts to H_2S[5].

The dry matter contents, the ammonia and the fatty acid concentrations are usually determined in a laboratory on a regular basis – ideally monthly.

When co-substrates are used to supplement the manure it is vital to analyse and control the characteristic parameters of the substrate, the biogas and the digested material in order to ensure smooth operation and high performance of the biogas plant.

The control unit of a biogas plant receives a set of measured parameters in order to support automated operation and determine the performance of the system. Among the parameters that are controlled by the control unit are the process temperature and the stirring unit.

3.2.4.13 GRID CONNECTION

In general cogeneration engines are used to supply the generated electricity to the electric grid. Running these engines in parallel to the electric grid requires that several technical rules need to be followed for safety reasons. Measurement devices, control units, switching and guard elements need to be installed in order to comply with electrical safety. Also, it is necessary to ensure that reactive power is compensated for, and that voltage fluctuations comply with local standards for the grid. Commonly they have to be within a range of ±3% to avoid disturbing sensitive electronic equipment. In cases where the local grid is too weak to take on the electricity produced by the cogeneration engine it is necessary to have a transformer substation installed that makes it possible to feed into the higher-voltage grid.

3.2.4.14 SAFETY EQUIPMENT

The handling of fuels such as biogas always requires compliance with safety rules in order to minimize the risk of an accident and to ensure safe operation of the biogas system. At the very least there are certain system safety components that must be installed, and accident prevention regulations that must be followed. An overview of the recommended safety equipment is shown in Figure 3.29. In some countries there are special safety regulations specifically for biogas systems (see also Chapter 8).

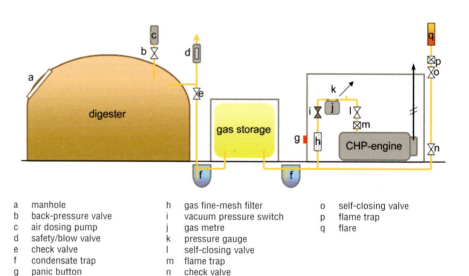

Figure 3.29.
Overview of biogas system safety components[5]
Graphic: Ecofys bv / www.ecofys.com

a	manhole	h	gas fine-mesh filter	o	self-closing valve
b	back-pressure valve	i	vacuum pressure switch	p	flame trap
c	air dosing pump	j	gas metre	q	flare
d	safety/blow valve	k	pressure gauge		
e	check valve	l	self-closing valve		
f	condensate trap	m	flame trap		
g	panic button	n	check valve		

As well as the safety equipment shown, there are additional rules to follow: for example, it is important to maintain the distance between the digester, the CHP engine and the stables or other buildings. There are also safety measures that can be addressed in the design of the biogas plant. For example, tank openings should be large enough to ensure sufficient ventilation; check valves and other safety switches must be easily reached; all the gas piping needs to be corrosion-resistant (copper piping does not fulfil this requirement); and the housing of the CHP engine needs to be adequately ventilated to provide a sufficient rate of air change.

*Figure 3.30.
Safety equipment
(flare, safety valve etc.)
Photo: A) Ecofys bv / www.ecofys.com
B) ABR Agrar Bio-Recycling GmbH /
www.abr-w.de C) PlanET GmbH /
www.planet-biogas.com D) Smack AG /
www.schmack-biogas.com*

3.3 Planning an anaerobic digestion project

3.3.1 Steps in project development

Project development can be regarded as the whole process that is required to realize an operational anaerobic digester. The various steps of this process are shown in Figure 3.31.

The process of project development starts with an idea that is outlined at a basic level in order to provide a general impression of the feasibility (project creation). This is then worked out in more detail to provide a detailed overview of the technical,

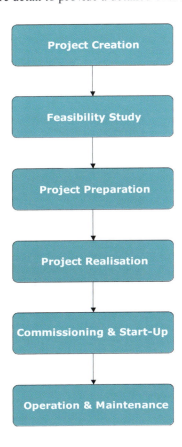

*Figure 3.31.
Schematic overview of
project development
Graphic: Ecofys bv / www.ecofys.com*

economical and legal feasibility of the project (feasibility study). If the feasibility looks promising, all the actions that are required to start the actual realization of the installation are undertaken (project preparation). At this point, the anaerobic digester can be constructed (project realization). After this phase the anaerobic digester is ready for commissioning and start-up. This section deals with all aspects of the development of an anaerobic digester up to project preparation. Project realization, and the commissioning and start-up, are discussed in section 3.4. Section 3.5 deals with the operation and maintenance of an anaerobic digester.

3.3.2 Project creation

3.3.2.1 INTRODUCTION

Biogas projects are complex, and their economic success depends on a variety of aspects that influence the technical and economical feasibility. It is therefore important to consider the relevant technical, organizational, economical and financing issues at an early stage when developing biogas projects. In this first phase, project creation, a number of relevant questions have to be answered positively:

- What kind of technique will be used? Is the existing infrastructure of the desired location (a farm, for example) used in an optimal way?
- What types of facility are required? For example, is it technically possible to feed in the electricity using the current grid connection? Can the produced heat be used?
- Is it possible to use co-substrates from within a close radius?
- How can the (co-)digested manure be disposed of?
- Is the project economically feasible?
- Is it likely that the necessary permits will be obtained?

Economic feasibility is the basis of every commercial project. In the creation phase, a cost analysis at a basic level is sufficient. In section 3.6.4 a calculation method is presented for estimating the economic feasibility of a digester. If this and the other questions have a positive outcome, the project development can continue with the next phase: a feasibility study, including a detailed economic analysis based on quotes from suppliers.

Figure 3.32.
A farm-scale anaerobic digester
Photo: Ecofys bv / www.ecofys.com

3.3.2.2 ANAEROBIC DIGESTION: ECONOMICALLY FEASIBLE?

The economic feasibility of AD depends on various factors. The feed-in tariff for electricity from biogas and possible subsidy schemes are determined mainly by government, but the following factors are farm, company or project dependent:

- Amount of usable manure. Increasing the amount of manure will lead to economies of scale: for example, doubling the amount of manure to be digested will increase the investment costs, but will not double them.
- Composition of the manure. The type of manure (from cattle or pigs, for example) determines the biogas yield, but the dry matter in the manure is also an important factor. If the manure is relatively wet, a large digester is required for a small biogas yield. In practice, the dry matter content can be influenced to some extent by reducing the water use when cleaning the stables.
- Availability of other digestible organic material (co-substrates). Co-digestion of other organic materials will increase the economic feasibility. Organic materials such as agricultural residues (possibly from own farm) or residues from the food processing industry have high specific biogas yields compared with manure. In most cases these residues can be obtained at low or negative costs.
- Consumption and cost of electricity use of own company/farm. The electricity that is produced can either be used on site by the farm or company, or fed into the electricity grid. This consideration will depend on the tariffs of both, but also on the sustainability goals of the owner of the biogas plant.
- Heat demand of own company or nearby neighbours. The heat produced with the biogas can satisfy a local heat demand. However, transport of heat is relatively expensive. So the final heat demand should be within a small radius of the heat production (a maximum of 200 m as a rule of thumb, but preferably less).
- End-use of (co)digested manure. The number of directly available nutrients is higher in (co)digested manure than in non-digested manure. It is possible to save more than 10% of the costs of nitrogenous fertilizer. But in practice this advantage is only possible when the substrate is used on one's own land. Buyers of manure are not likely to pay more for it to be digested.
- Financial status. It is advisable to think about the financing of the project at an early stage. If the creditworthiness of the investor is high the interest rate of the loan required to finance the investment can be lower. Also, the amount of the investment that can be obtained by means of a loan will be higher. One possible option for financing a project is through an external investor, such as a leasing company or an electricity company.

3.3.2.3 REQUIRED PERMITS

In general, building permits and environmental permits are necessary for AD installations. It is advisable to check whether permission is needed to use the (co)digestate as a fertilizer. It should also be checked whether there is a zoning plan for the desired location of the digester.

For each permit it is important to check with the respective legal authority on their general attitude towards AD, and how long the permission process will take. The relevant regulations vary from country to country. A detailed description of these and other required permits is given in Chapter 8.

3.3.2.4 GRID CONNECTION

In order to feed the produced electricity into the grid, adaptations of the existing connection are usually required. In the creation phase of planning it is usually sufficient to know what types of extra facility will be necessary if a small CHP unit (20–150 kW) is to be connected to the grid. The owner of the grid can provide information on the conditions that the installation must fulfil, and on the costs related to the necessary adaptation of the grid. It is advisable to contact the local grid operator at an early stage of the project. Usually the electrical equipment and the grid connection have to be executed by a registered electrician.

3.3.2.5 GO/NO GO

At the end of the creation phase one should be able to give a positive answer to the questions stated at the beginning of this phase. A first impression of the desired installation, including capacity, and the outline of the feasibility has been sketched. At this point the project can move to the next phase: the feasibility study.

3.3.3 Feasibility study

3.3.3.1 TENDERS FROM SUPPLIERS

There are many suppliers of biogas installations. Requesting tenders is a good way to get an idea of the differences in technique and investment costs between the various suppliers. A tender can be turnkey, offering an operational installation. Also, activities that are required to operate an anaerobic digester legally may be included. Alternatively, some of these things can be done on one's own account. Often parts of the construction, obtaining permissions and making arrangements with the grid operator are done by the farmer or company who wants to set up the anaerobic digester.

When requesting a tender it is important to be clear what type of installation is wanted and what should be included or excluded. This can be spelled out in a specifications, which should include the following minimum requirements:

- annual amount of manure
- composition of the manure (or at least the dry matter fraction)
- annual amount of co-substrates, physical description and composition
- existing infrastructure that can be integrated with the AD installation, including manure storage systems that can be used for storage of the (co)digested manure, and existing buildings that can be used or existing equipment such as manure mixers or pumps
- description of the soil or the foundation on which the digester will be installed
- heat demand that will be satisfied by the heat from the CHP, such as the stables, the (private) house or other.

It is also clear that the installation needs to obey the applicable standards regarding safety, emissions and noise. The supplier should be familiar with these standards.

On the basis of the main components of a digester and their size it is also possible to form one's own estimate of the investment costs for an installation. The sizing of the components is worked out in section 3.3.3.7.

3.3.3.2 PERMITS

In the creation phase it was sufficient to find out about the general attitude of the legal authority towards AD. In this phase a further step has to be taken and a request for a decision in principle has to be passed on. It is likely that a small description of the project with a sketch of the intended result will be sufficient.

3.3.3.3 FEED-IN OF ELECTRICITY

Usually, most of the electricity will be fed into the grid, as this is economically more attractive. However, at some times – at peak hours, for example – it can be more favourable to use the produced electricity oneself. Also, in some cases it can be advantageous to have a larger CHP unit in order to produce electricity at peak hours only. This decision should be made on basis of the peak and off-peak tariffs, the feed-in tariffs, and the additional costs for the CHP unit. The electricity company involved could give advice in making this decision.

3.3.3.4 UTILIZATION OF HEAT

In the creation phase an inventory was made of the possible use of the produced heat. In this phase the heat demand must be worked out in detail. As well as the size of the heat demand, its variation in time is important: for example, for households the heat demand may be nearly absent in the summer. Also, the benefits of the heat use must be weighed against the costs of heat pipes (Figure 3.33).

*Figure 3.33.
Example of heat piping
Photo: ABR Agrar Bio-Recycling GmbH
/ www.abr-w.de*

3.3.3.5 CONTRACTING FOR CO-SUBSTRATES

If the farm cannot supply the desired amount of co-substrates (or manure) itself, external supply is necessary. In some cases this supply will be done on the basis of accidentally available residues. However, for a continuous supply of co-substrates it is advisable to conclude contracts with these suppliers. Important aspects that should be arranged in such a contract are:

- type of co-substrate (or manure) and amount
- rate(s) and time of supply
- specifications of quality (possibly in ranges), including moisture content, contamination (e.g. plastics, stones), nutrients
- measurement of the deliverable (quantity and quality). Which (standard) methods will you use for this? On what basis can you refuse a load? Assign a procedure that deals with these steps
- duration of the contract. The longer the contract, the more continuity of the supply is secured. Also set down a period of notice

*Figure 3.34.
Inserting co-substrates into the digester tank
Photo: PlanET GmbH / www.planet-biogas.com*

- price of the co-substrate. Prices can fluctuate annually: therefore it might be necessary to negotiate a price every year
- conditions of payment
- liability.

3.3.3.6 DISPOSAL OF EXTRA NUTRIENTS

The extra nutrients resulting from the supply of co-substrate from external sources have to be used after digestion. If this can be done on one's own land it may not result in extra costs (apart from extra costs for transport to and applying it on the land). However, if there is no space left on one's own land the extra nutrients will have to be bought by a third party. Additional contracts for this will be necessary.

3.3.3.7 SIZING

In the tender from the supplier the size of the various components should be specified. On the basis of the main components of a digester and their size, one's own estimate of the investment costs for the installation should be made. The main cost components are the digester tank and its insulation, the CHP unit, the mixers, the pumps and the piping. As a rule of thumb the following formulae can be used to calculate the required size or volume of the various components. The same example – the digestion of 5000 m³ of cattle manure and 1000 m³ (800 t) of agricultural waste per year – is used for all example calculations.

3.3.3.7.1 DIGESTER VOLUME

$$\text{Digester volume (m}^3\text{)} = \left[\text{manure (m}^3\text{/yr)} + \text{co-substrate (m}^3\text{/yr)} \right] \times \frac{\text{Retention time (days)}}{365}$$

For mesophilic digestion the retention time will be around 30 days.

Example
5000 m³ of cattle manure and 1000 m³ of agricultural waste are annually digested with a retention time of 28 days. The digester volume will need to be at least

$(5000 + 1000) \times (28/365) = 461$ m³

3.3.3.7.2 POST-DIGESTION STORAGE

In many cases it is practical or required to store the digestate. In most manure cellars with (semi)-open floors it is not practical to separate the fresh manure from the digestate. In this case external storage for the digestate is required. This can be existing storage (such as a silo or a manure bag) or a new storage. The size of this storage can be calculated as follows:

$$\text{Size of storage (m}^3\text{)} = \text{Animal input substrate (m}^3\text{/yr)} \times \frac{\text{Required storage time (months)}}{12} - \text{Size of digester (m}^3\text{)}$$

Example
5000 m³ of cattle manure and 1000 m³ of agricultural waste are annually digested with a retention time of 28 days. A storage time of 2 months is required. The size of the post-digestion storage is

$(5000 + 1000) \times (2/12) - 461 = 539 \text{ m}^3$

3.3.3.7.3 BIOGAS PRODUCTION

The biogas production is determined by the dry matter content (DM), the organic fraction of the dry matter (OM/DM), and the biogas production per kg of OM (see section 3.2.2.4). The following formula can be used to calculate the biogas production:

$$\text{Biogas production (m}^3\text{/yr)} = \left[\text{Manure}_m \text{ (t/yr)} \times DM_m \times \frac{OM_m}{DM_{cs}} \times (\text{m}^3 \text{ biogas/kg OM}_m) \times 1000\right] + \left[\text{Co-substrate} \times DM_{cs} \times \frac{OMcs}{DM_{cs}} \times (\text{m}^3 \text{ biogas/kg OMcs}) \times 1000\right]$$

Example
5000 m³ of cattle manure and 1000 m³ of agricultural waste are digested annually. The cattle manure (with a density of 1 t/m³) has a DM of 10%, an OM/DM of 80% and a biogas yield of 0.25 m³/kg Om. The organic waste (with a density of 0.8 t/m³) has a DM of 30%, an OM/DM of 70% and a biogas yield of 0.55 m³/kg Om.

The biogas production (m³/yr) = [(5000 × 1) (t manure/yr) × 10% × 80% × 0.25 × 1000] + [(1000 × 0.8) (t waste/yr) × 30% × 70% × 0.55 × 1000]

= 100,000 + 92,400

= 192,400 m³/yr.

3.3.3.7.4 BIOGAS STORAGE

The biogas storage is done either in an external gas bag or by means of a foil that covers the silo. In practice a storage capacity of 20–50% of the daily biogas production is sufficient for use in a CHP unit. This might be even less if the CHP unit is constantly operational.

For an external gas bag:

Size of biogas storage (m³) = Daily biogas production (m³/day) × 20%

Example
A biogas production of 192,400 m³/yr means 527 m³/day. This requires a biogas storage of

527 × 20% = 106 m³

For biogas foil, the size of the required foil to cover the silo is determined by the diameter of the digester tank. The amount of gas stored under the foil is relatively small. This can increase if the digester is not entirely filled, as all excess volume can be used for gas storage. In practice it might be necessary to use a slightly larger digester to compensate for this loss in storage capacity.

$$\text{Diameter of digester (m)} = 2 \times \sqrt{\frac{\text{Volume of digester (m}^3\text{)}}{\text{Height of digester (m)} \times 3.14}}$$

Example
A digester of 461 m³ capacity is 5 m high. The diameter of the digester is therefore equal to

$$2 \times \sqrt{\frac{461}{5 \times 3.14}} = 2 \times \sqrt{\frac{92.2}{3.14}} = 2 \times 5.4 = 10.8 \text{ m}$$

In practice it might be useful to choose a bigger digester in order to increase the biogas storage capacity.

3.3.3.7.5 CHP CAPACITY

$$\text{CHP capacity (kW}_e\text{)} = \frac{\text{Biogas production (m}^3\text{/yr)} \times [\text{ Calorific value of biogas (MJ/Nm}^3\text{/3.6]}}{\text{Operational full load (h/yr)} \times \text{Electrical efficiency}}$$

The calorific value of biogas (in MJ) can be calculated as: methane content biogas × 34. An average value is 20 MJ/Nm³. As a rule of thumb an electrical efficiency of 30% is used. For CHPs bigger than 50 kW this may increase; for CHPs smaller than 30 kW it may decrease. If the CHP unit is used full-time the number of operational hours will be around 7500 per year.

Example
5000 m³ of cattle manure and 1000 m³ of agricultural waste are digested annually and produce 192,400 m³ biogas/yr. The CHP unit required is:

$$\frac{192{,}400 \times 20/3.6}{7500 \times 30\%} = 42.8 \text{ kW}_e$$

The following formulae are useful for sizing the flare and heating components:

$$\text{Thermal input CHP (kW}_{th}\text{)} = \frac{\text{CHP capacity (kW}_e\text{)}}{\text{Electrical efficiency}}$$

$$\text{Thermal output CHP (kW}_{th}\text{)} = \text{Thermal input CHP (kW}_{th}\text{)} \times \text{Thermal efficiency CHP}$$

An average CHP unit for farm-scale digesters has a thermal efficiency of about 50%.

Example
The 42.8 kW$_e$ CHP unit has a thermal input of 42.8/30% = 142.7 kW$_{th}$. The thermal output is equal to 142.7 × 50% = 71.4 kW$_{th}$.

3.3.3.7.6 INSULATION MATERIAL
The insulation for the walls of the digester is given by:

Area of wall insulation (m) = Height of digester (m) × diameter (m) × 3.14

Example
A digester of 461 m³ is 5 m high. The diameter is 10.8 m. The area of the insulation is:

5 × 10.8 × 3.14 = 170 m²

When the insulation is 6 cm thick, the volume is

Area × Thickness = 170 × 0.06 = 10.2 m³

In some cases the bottom of the digester will also be insulated. If this is done, the following formula applies:

Area of bottom = Diameter² × 0.785

Example
A digester of 461 m³ is 5 m high. The diameter is 10.8 m. The area of the insulation of the bottom is

10.8² × 0.785 = 91.6 m²

When the insulation is 8 cm thick, the volume is

Area × Thickness = 91.6 × 0.08 = 7.3 m³

3.3.3.7.7 HEAT PIPING FOR RESIDUAL HEAT
Heat demand of the digester
A large part of the heat produced is used for maintaining the temperature in the digester. Therefore heat is required to warm up the fresh substrate, and to compensate for energy losses through transmission. The latter depends on the isolation of the digester and the temperature outside the digester. As a rule of thumb, this is about 30% of the energy required for the heating of the substrate. The amount of heat required to maintain the temperature in the digester can be calculated as following:

$$\frac{\text{Heat demand}}{(\text{MJ/yr})} = \frac{\text{Mass of substrate}}{(\text{t/yr})} \times \frac{\text{Specific heat}}{(\text{KJ/kg/K})} \times (\text{T digester} - \text{T fresh substrate}) \times 130\%$$

As a rule of thumb the specific heat of the substrate is equal to that of water (4.2 kJ/kg/K). For substrates with a relative low water content the specific heat will be lower.

Example
5000 t of cattle manure and 800 t of agricultural waste are digested annually at a (mesophilic) temperature of 35°C. The average temperature of the fresh substrate is 15°C. The heat demand of the digester is equal to:

(5000 + 800) × 4.2 × (35 – 15) × 130% = 633,360 MJ/yr

This equals 633 GJ/yr.

Note: When sanitation is applied, the (co-)substrate is pre- or post-heated to a higher temperature. The specific additional heat demand for this process depends largely on the configuration (for example using heat recovery, heating only the co-substrate, pre-

or post-heating). The additional heat demand will be in the range of 10–140% of the heat that is used in the digester.

Residual heat
The residual heat is the heat that is still left when the heat demand of the digester is subtracted from the total heat production of the CHP unit. This heat can be used effectively, for example to heat the stables or a private house.

As formulae:

$$\text{Heat production of CHP (GJ)} = \text{Thermal output of CHP (kW}_{th}\text{)} \times \text{Operational hours of CHP}$$

$$\text{Residual heat (GJ)} = \text{Heat production of CHP (GJ)} - \text{Heat demand of digester (GJ)}$$

Example
5000 t of cattle manure and 800 t of agricultural waste are digested annually. The thermal (output) capacity of the CHP unit is 71.4 kW$_{th}$, and it operates for 7500 h/yr. The heat production of the CHP unit is

71.4 × 7500 × 3.6/1000 = 1.928 GJ/yr

The residual heat is

1928–633 = 1295 GJ/yr

Sizing of the heat piping
If the residual heat production and the heat demand of buildings etc. is known, it is possible to assess whether it is economically attractive to use it (see also section 3.6). In most cases the heat pipes will be sized on the basis of the full CHP heat production capacity. In this way there is a degree of flexibility in the division of the heat between the digester and the buildings etc. For example, at certain times all of the produced heat may go to the buildings. For farm-scale digesters a heat pipe with a diameter of 33.7 mm and a thickness of 2.6 mm will be more than sufficient.

Required capacity of heat pipes (kW$_{th}$) = Thermal of output CHP

Table 3.6 shows the required size of the heat pipes for various capacities.

Table 3.6.
Required size of heat pipes for various capacities

Maximum capacity (kW$_{th}$)	Diameter (mm)
18	13.5
30	17.2
45	21.3
70	26.9
110	33.7
175	42.4

Example
5000 t of cattle manure and 800 t of agricultural waste are digested annually. The thermal (output) capacity of the CHP unit is 71.4 kW$_{th}$. The required heat pipe has a minimum diameter of 26.9 mm (or 33.7).

3.3.3.7.8 MANURE PUMP
The type and size of the manure pump depend on the amount of slurry and its dry matter content of the slurry, and on the height to which the slurry has to be pumped (the entry point of the digester).

Example
A 460 m^3 digester tank receives, three times a day, 5 m^3 manure over an hour's time. The manure has a dry matter content of 7–10%. A vane pump of 3 kW will be sufficient.

3.3.3.7.9 MIXER

The type and size of the mixer depend largely on the dry matter content in the digester, and on the size of the digester tank. The capacity of the mixer will be in the range 2–25 kW.

Example
A 460 m^3 digester tank contains slurry with a dry matter content of 7%. A plunge mixer of 7.5 kW is required. If the dry matter content increases to 10%, a plunge mixer of 11 kW will be required.

3.3.3.8 GO/NO GO

With the information acquired in the feasibility study, the economic feasibility can be estimated using the calculation tool in section 3.6.4. If the economic feasibility is positive, and if the legal authority has given a positive decision in principle on the proposed anaerobic digester, the project can be worked out in more detail.

3.3.4 Project preparation

3.3.4.1 SELECTION OF SUPPLIER

On the basis of the various tenders (from the feasibility study) a preferred supplier can be selected. With this supplier (or with multiple suppliers for the various parts of the digester), the terms of delivery have to be agreed. The following aspects are important to consider:

- What exactly will be delivered? For example, what is the size of the digester? Are the permits included? Are there possible additional costs?
- What type of effort is required during construction (amount of time, required skills)? Costs can be reduced if the project owner provides labour to help during the construction of the installation.
- What is the delivery period?
- Product and process specifications
- Guarantees for the project or the process (for example duration of guarantee, minimum number of operational hours, minimum yields)
- The price of the offer, the period for which the offer is valid, and the index that can be used to adjust the price.
- Terminating conditions: reasons to dissolve the contract. These should at least include the refusal of permits (or additional requirements by the legal authority that are unacceptably costly) and the failure of the project financing.
- Possibilities for a maintenance contract. What exactly is included in the maintenance contract? It is advisable to have at least a maintenance contract for the CHP unit.

At this stage the supplier will have to detail out the installation (engineering). Specifications, a scale drawing, maps etc. are likely to be necessary for the permissions. The supplier may ask for reimbursement for this engineering. This can be deducted when the installation is commissioned.

3.3.4.2 PERMISSIONS

At this stage the permissions process can be started. The legal authority will tell you which documents are needed (see Chapter 8). In most cases the minimum requirements will be:

- scale drawings
- engineering calculations (e.g. mass flows, biogas and kWh production, sound levels)
- safety plan.

It is possible that there will be additional requirements. If these have high costs associated with them, the economic feasibility of the project can be endangered. For this reason, the plant should not be commissioned until the permissions have been obtained.

3.3.4.3 PROJECT FINANCING

During the permissions process the project financing can be worked out in detail. Possible subsidy schemes should be checked (see Chapter 9). As in most cases a loan will be necessary, it is wise to ask various financing institutions (banks or leasing companies) for quotes. An accountant or a financial/legal advisor can give advice on the optimal legal structure (e.g. a corporation).

3.3.4.4 FINAL GO/NO GO

On the basis of the detailed design of the installation, meeting all the requirements set out in the granted permits, the final economic analysis can be made. This analysis should show the annual cashflow over the lifetime of the project. As the project financier will check the cashflow, it is advisable to let an accountant control the calculations. If this final economic analysis is positive the project is ready for realization. Nonetheless, the following checklist should be considered first:

- There are no misunderstandings between the supplier and the customer. It is clear what the supplier will deliver and what your effort will be.
- The building site is reachable, the soil conditions are OK, and the positions of existing wires and cables are known.
- All permits are irrevocably granted.
- All tasks are commissioned by written documents.
- There is a working plan for the installation, including the connection of the separate parts.
- Payment conditions are agreed upon with all the involved parties.
- Approved construction plans are present.
- The quality plan is present.
- The building plan is present

3.4 Project realization, commissioning and start-up

Once the final go decision has been made, the project is ready for construction of the anaerobic digester. In this section the planning of the construction up to the start-up of the installation is discussed.

3.4.1 Planning and construction

3.4.1.1 PLANNING

The companies that are responsible for the construction and the installation must have a clear plan containing the following data:

- the start of the work, and the time planning in phases with corresponding milestones/deliverables
- the supply of parts and materials
- payments
- completion and completion test.

The communication structure must also be clear beforehand. When and with whom will consultations take place, in order to discuss the progress and possible difficulties? Is there a building supervisor who should see to it that the installation is in compliance with the specifications and the legal requirements? There should also be a procedure to notice extra or less work at an early stage.

3.4.1.2 CONSTRUCTION

During the construction (Figure 3.35) the following things should be well documented:

- specifications of the built installation (with possible differences from the original specifications)
- test results
- instructions for operation and maintenance and safety procedures (training of personnel)
- results of the commissioning test
- calculation of the realized investment costs
- guarantees and quality certificates.

Figure 3.35.
AD plant under construction
Photo: Krieg & Fischer Ingenieure
GmbH / www.kriegfischer.de

3.4.1.3 OWN CONTRIBUTION DURING CONSTRUCTION AND SUPERVISION

It is possible that the project owner will have agreed with the supplier to help during construction. Assistance in several activities is possible, such as insulating the digester tank, mixing cement and pouring concrete, unloading equipment, welding or gluing together of the digester tank, connecting pipes, wires etc. It is also advised that the project owner does a regular check of the progress of the construction.

3.4.2 Start-up

The start-up of the anaerobic digester is a crucial step in the project realization. During start-up the biological process of biogas production will commence. The bacteria that are responsible for this process are already present in cattle manure, but need to be added when pig manure is used. Over a period of 3 months the biogas production will gradually increase to its maximum. The composition of the produced biogas can fluctuate during this start-up period. The methane content will increase up to 55–60%. The sulphur concentration in the biogas is high at the beginning, but will decrease once the (biological) desulphurization is operational.

When the construction of the digester is completed, it is advisable that the supplier supervises the start-up. After the start-up period the buyer can check whether the installation meets the specifications regarding the biogas yield and electricity production. The day-to-day activities that the operator of the plant will have to perform during start-up consist of:

- feeding-in of manure or, if this is done automatically, the monitoring of this
- adding the co-digestates
- monitoring the function of the mixers
- keeping a logbook with the daily inputs of manure and co-digestates, temperature in the digester, biogas yield etc.

It is important that the tasks of the operator are well documented. During start-up the supplier should still be responsible.

When the methane content of the biogas is below 45% there may be a risk of explosion. If the methane content is higher than 45% the gas will burn without a pilot flame. The following safety precautions should be taken during the start-up period:

- Prevention of sparks.
- Disconnection of the gas conversion equipment from the digester.
- During start-up the substrate has to be heated. As there is, as yet, no biogas to fuel the CHP unit, an alternative fuel or heat source is necessary.
- If the CHP unit is connected to the heating system of, for example, a farm or a company, existing boilers can be used.
- If the CHP unit offers the possibility of operating in a dual-fuel mode (for example, if it can operate on either biogas or diesel), the secondary fuel can be used to fire the CHP unit.
- A burner fuelled by diesel, natural gas, propane or another fossil fuel can be used temporarily.

When fossil fuels are also being used to fire the CHP unit, the produced electricity can't be regarded as fully sustainable. If this electricity is to be fed into the grid, discuss this issue with the electricity company that will buy it.

In some cases where organic material is co-digested, it can be possible that the environmental permit requires samples of the co-digested material. It is advisable to have samples analysed already during the start-up of the digester. It should be the responsibility of the supplier of the digester to ensure that the composition of the input and output of the anaerobic digester during this start-up period complies with the regulations.

In most cases the supplier will hand over responsibility for the operation of the anaerobic digester to the buyer at the end of the start-up period. Therefore the buyer of the digester will have to ensure during start-up that the installation is operating in accordance with the guaranteed specifications, such as the biogas yield and the composition.

The supplier must inform the user adequately about the operational aspects. These include:

- instruction on the daily routines (feeding of manure and/or co-digestates, mixing)
- inspection of the process's main parameters and indicators (reading out measuring equipment)
- monitoring of the biogas yield and composition (sulphur content, methane content)
- operation and maintenance of the biogas conversion equipment (CHP, burners, flare)
- safety instructions: indication of explosion alarm, measures in case of exceeding limiting values and emergencies
- monitoring and administration (possibly as a legal requirement)
- settling of accounts with the electricity company regarding the electricity that is fed into the grid.

As a result of this instruction the user should be capable of operating and maintaining the anaerobic digester.

3.5 Operation and maintenance

After start-up of the digester, it has to be operated and maintained. This section deals with the operational aspects of an anaerobic digester, under normal conditions and in the case of a malfunction, as well as with the maintenance aspects.

3.5.1 Operation of the digester under normal conditions

The bacteria in the digester will take care of the biogas production. It is the task of the operator to control the process conditions and to make sure that the CHP unit operates properly. In order to do so the following activities will have to be undertaken.

3.5.1.1 DAILY ACTIVITIES
- inputting the manure and co-substrates into the digester tank
- inspecting the engine oil
- checking the fault display/lights of the switchboard cabinet
- inspecting the water pressure of the heating equipment
- inspecting the dosing pump of the desulphurization unit
- monitoring the temperature in the digester tank
- adjusting the mixing intervals in order to avoid a floating surface or deposition on the bottom; also ensuring that the interval allows the biogas to escape from the slurry gradually
- inspecting all supply and drainage pipes for flow of the manure and co-substrates
- inspecting the levels in the digester tank and in the final storage tank
- inspecting the biogas storage
- recording the biogas yield and operational hours of the CHP unit.

Other relevant aspects that should be registered are the biogas consumption of the CHP unit, the electricity production, the digestion temperature, the input of co-substrate, maintenance activities, and any unusual incidents.

These activities will take about 30 mins per day. If co-substrates are fed into the digester manually, more time may be necessary.

3.5.1.2 WEEKLY ACTIVITIES
- Check the levels in the bags containing water from condensation, and empty if necessary.
- Test the mixers.
- Visually check the CHP unit and all its pipes.
- Check the overpressure valve for functioning and pollution.

3.5.1.3 HALF-YEARLY ACTIVITIES
- Inspect all bolts and flaps.
- Bleed the central heating.
- Inspect all electric equipment for damage.
- Inspect underpressure safeguards.
- Inspect all safety equipment.

3.5.1.4 YEARLY ACTIVITIES
- Inspect the part of the installation that contains biogas for damage, leakage and corrosion.
- Test the fire extinguisher(s).
- Check all liquids for frost resistance.

3.5.1.5 OTHER ACTIVITIES
- Accept co-substrates from external supplier (taking samples if necessary).
- Financial administration.

3.5.2 Operation of the digester during malfunction

AD is a proven technology. However, it is possible that a malfunction will occur. In most cases a maintenance mechanic will have to come to fix the problem. Below is described what actions should be taken or what risks are present in the case of malfunctioning of parts of the installation.

If the biogas storage malfunctions, the required actions are:

- Cut off the gas supply to the biogas storage.
- Empty the biogas storage.
- Enter the storage only after sufficient ventilation, and in the presence of a second person who is holding a lifeline.

Malfunctioning of the heating system: a leak in the heating system or its piping induces the risk of burns from the heat transfer fluid.

If the CHP unit malfunctions, the required actions are:

- Cut off the gas supply outside the CHP unit/building.
- Press the emergency stop outside the CHP unit/building.
- In case of gas odour, ventilate, and avoid sparks or open fire.

If the electronics default, this should be solved by a professional electrician/expert.

If there is a malfunction in the manure pipes, pumps or mixer, the required actions are:

- Remove blockages immediately.
- If the malfunction is in a pump or mixer, close all flaps, and switch off and secure all pumps.

If the digester tank malfunctions, when entering the tank there should be sufficient ventilation. If not there is a risk of suffocation, poisoning, fire or explosion.

3.5.3 Maintenance

The various components of an anaerobic digester can be subject to malfunctions, as described in the previous section, and in all cases they will be subject to wear. Therefore maintenance is required periodically (Figure 3.36). The CHP unit is serviced every 20,000 operational hours and needs an overhaul every 60,000 operational hours. The pumps (mainly the vanes) need to be overhauled every 3–5 years. Under normal circumstances the other components of the installation are not likely to wear out before the end of the technical lifetime (10–20 years), but some repairs may be necessary. The operator of the biogas plant can perform simple repairs, such as the unblocking of a pipe or the replacement of a bolt. However, for more complex maintenance, such as the maintenance of the electric equipment, the replacement of a biogas pipe or the repair of a pump, it is advisable to have the help of a maintenance craftsman/expert.

Figure 3.36.
Maintenance activity
Photo: Ecofys bv / www.ecofys.com

3.6 Economics

3.6.1 Introduction

As for all investments, the economic feasibility is an important factor in the final go/no go decision. In this section the costs and benefits of an anaerobic digester are discussed. With the information in this section it is possible to make a first-order estimate of the costs (including an estimate of the investment costs) and benefits for a specific situation. For a further level of detail, as required for the feasibility study, it is advisable to make use of tenders from the various suppliers. At the end of the section a calculation tool that can be used to estimate the economic feasibility is presented, together with some examples.

3.6.2 Costs

3.6.2.1 COST COMPONENTS

The various cost components for an operational digester are:

- investment costs
- operation and maintenance costs
- insurance and taxes
- intake and end-use of co-digested manure.

3.6.2.2 INVESTMENT COSTS

The investment costs for an anaerobic digester will vary from case to case, depending on the specific needs of the installation. As a result it is difficult to specify investment costs beforehand. Total investment costs for an anaerobic digester may vary from €2500 to €7500 per or from €250 to €700 per m^3 of digester.

The investment cost for a specific case can be estimated by adding up the costs of the main components of the installation, including costs for engineering and project development. The size of these components can be determined using the formulae in section 3.3. An overview of the main components of the installation and their costs is presented below. These cost figures are indicative only, and they can vary from country to country.

3.6.2.2.1 DIGESTER TANK

In Figure 3.37 the low costs refer to a concrete silo with a concrete cover, excluding installation costs. In practice it is possible to perform the installation on one's own. The high costs refer to an installed concrete silo with a pyramid-shaped synthetic cover. An enamelled steel silo will be slightly more expensive.

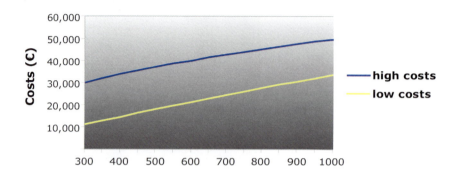

Figure 3.37.
Investment costs for a digester tank
Graphic: Ecofys bv / www.ecofys.com

3.6.2.2.2 POST-DIGESTION TANK

See Figure 3.38. The costs for the manure bag exclude the cost of soil preparation, which may be necessary.

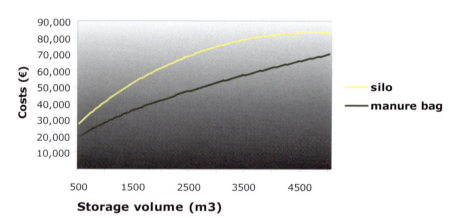

Figure 3.38.
Investment costs for a manure bag
Graphic: Ecofys bv / www.ecofys.com

3.6.2.2.3 BIOGAS STORAGE

The produced biogas can be stored either in an external gas bag (Figure 3.39) or 'on top' of the manure, using a silo-covering foil (Figure 3.40).

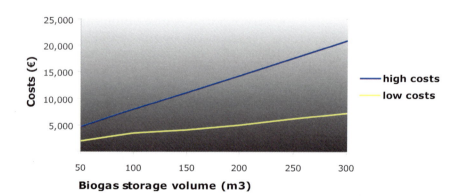

Figure 3.39.
Investment costs for a gas bag
Graphic: Ecofys bv / www.ecofys.com

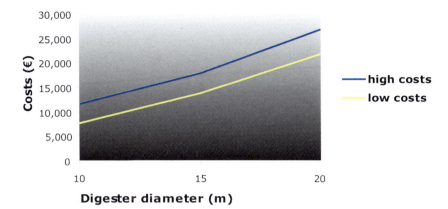

*Figure 3.40.
Investment costs for a silo-covering foil, excluding any possible construction required for the support of the foil
Graphic: Ecofys bv / www.ecofys.com*

3.6.2.2.4 CHP UNIT

In Figure 3.41 a large range in investment costs for CHP units is shown. The high costs refer to a CHP unit including accessories such as sound-insulating case, lambda control, switch box, condenser and flame control. The low costs include only a heat exchanger and an emergency cooler.

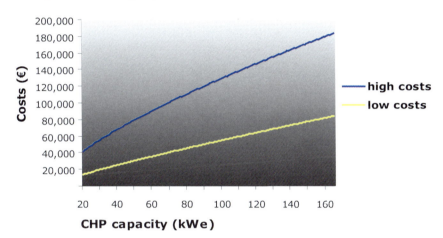

*Figure 3.41.
Investment costs for a CHP unit
Graphic: Ecofys bv / www.ecofys.com*

3.6.2.2.5 INSULATION

- Polyurethane (PUR, commonly used for sides of digester): €500–800/m^3, 0.06 thick.
- Polystyrene (commonly used for bottom of digester): €175–275/m^3, 0.08 thick.
- Mineral wool: €50/m^3, 0.1 m thick.

3.6.2.2.6 HEAT PIPES

The costs for installed return pipes are listed in Table 3.7. All costs mentioned are without improvements of the soil (e.g. cement or pavement) or other unforeseen objects. These cost items do not occur in most cases.

*Table 3.7.
Indication of investment costs in heat (return) pipes for supply of heat to external sources*

Inside diameter (mm)	Cost (€/m)
13.5	54
17.2	61
21.3	68
26.9	75
33.7	82
42.4	90

3.6.2.2.7 CONNECTION OF CHP ENGINE TO CENTRAL HEATING SYSTEM

- 20 kW$_{th}$ (thermal output of CHP): €1500.
- 50 kW$_{th}$: €3000.
- > 50 kW$_{th}$: €5000.

3.6.2.2.8 MANURE PIPES
- 50 mm × 2 mm (diameter × wall thickness), PN 7.5 (maximum pressure in bar): €1.60/m.
- 90 mm × 2.7 mm, PN 7.5: €3.75/m.
- 125 mm × 3.7 mm, PN 7.5: €6.1/m.
- 125 mm × 4.8 mm, PN 10: €9.1/m.

Installation costs: €3–15/m (depending on soil type, total length, distance to installation etc.).

3.6.2.2.9 PUMPS
- Vane pump (3 kW): €3000.
- Press pump (5.5 kW): €2800.

3.6.2.2.10 MIXERS
- Plunge mixer (7.5 kW): €5800.
- Plunge mixer (11 kW): €6800.

3.6.2.2.11 FLARE
- 30 kW_{th} (total input CHP): €3000.
- 60 kW_{th}: €5000.
- 100–200 kW_{th}: €10,000.

3.6.2.2.12 HEAT EXCHANGER INSIDE THE DIGESTER
- 20 kW_{th} (thermal output CHP): €5500.
- 30 kW_{th}: €7200.
- 40 kW_{th}: €8600.
- 50 kW_{th}: €10,000.
- > 50 kW_{th}: €15,000.

3.6.2.2.13 OTHER EQUIPMENT (INCLUDING SAFETY MEASURES)
- Desulphurization: €350.
- Condensate trap (for drying of biogas): €6000–10,000.
- Other accessories including H_2S detector, CO_2 indicator, leak detector, flame extinguisher: total costs for a farm-scale plant approximately €750.

3.6.2.2.14 CO-DIGESTION
If the installation will also digest co-substrates, there are additional costs for a larger size of the CHP unit, the digester tank, the gas storage and the post-digestion storage, and for stronger pumps and mixers (because of the higher dry matter content). The costs for the total installation can still be estimated using the figures above. However, there is some additional equipment required.

Input of co-substrate:

- Pre-mixing well: €25,000–35,000.
- Dry matter input: €35,000–45,000.

Fluid matter input (including storage tank):

- Without preheating: €2000–5000.
- Including preheating: €8000–15,000.
- Sanitation tank: €30,000 (for up to 9000 m^3/yr)

Storage of co-substrate: €0–10,000. Note: This depends largely on the type of co-substrate (e.g. fats or verge grass) and the available infrastructure (e.g. a shed, or a concrete floor).

3.6.2.2.15 CIVIL WORKS
This includes the preparation of the construction site, buildings etc., and differs for each specific situation. For example, the buyer can do the preparation of the

construction site himself, but in other cases a concrete floor may be necessary. The costs for this may vary from €0 to €20,000 for digesters up to 500 m^3, and may be higher for larger installations.

3.6.2.2.16 ENGINEERING AND CONSTRUCTION

Apart from the single components there will also be costs for the engineering and construction. Normal costs for this would be around 10% of the hardware costs. This percentage can be reduced if the installation is built under the buyer's own supervision or with a relatively large contribution by the buyer.

3.6.2.2.16 PROJECT DEVELOPMENT

This includes the permissions, the finalising of contracts etc., and may also include a feasibility study by an external agency. Costs may vary with the complexity of this process, and the buyer may carry out parts (or all) of the project development. For a standardised farm-scale anaerobic digester the costs for project development may be €5000 or less. However, for a relatively innovative system (such as a system using a 'new' co-substrate) the costs for project development may be €15,000 or more.

3.6.2.3 EXAMPLE CALCULATION OF INVESTMENT COSTS

Table 3.8 shows an example calculation of the investment costs for the digestion of 5000 m^3 of cattle manure and 1000 m^3 of agricultural waste. See section 3.3.3.7 for the sizing of the components for this example.

Table 3.8. Example calculation of the investment costs for the digestion of 5000 m^3 of cattle manure and 1000 m^3 of agricultural waste

	Size	Unit	Low costs (€)	High costs (€)
Digester	460	m^3	17,000	36,000
Post-digestion storage	539	m^3	20,500	29,500
Biogas storage	106	m^3	3500	8500
CHP unit	43	kW	27,000	73,000
Insulation: side (polyurethane)	10.2	m^3	5100	8160
Insulation: bottom (polystyrene)	7.3	m^3	1278	2008
Heat pipes	26.9	mm (thick)	7500	7500
	100	m (long)		
Connection to central heating	72	kW$_{th}$	5000	5000
Manure pipes	90 × 2, PN 7.5			
	30	m	203	563
Pump	Vane 3 kW		3000	3000
Mixer	Plunge 11 kW		6800	6800
Flare	143	kW$_{th}$	0	10,000
Heating of digester	72	kW$_{th}$	15,000	15,000
Other equipment and safety			7100	11,100
Pre-mixing well			25,000	35,000
Sanitation tank	Not included		0	0
Storage of co-substrate			0	10,000
Civil works			0	15,000
Subtotal			143,980	276,130
Engineering	5–10%		7199	27,613
Total costs of installation			151,179	303,743
Project development			5000	15,000
Total project costs			156,179	318,743

For this specific situation the project costs will be between €155,000 and €320,000. As an estimate the average of these two (€237,500) can be used, but in practice both extremes are possible. Note that if the post-digestion storage is not be necessary, for example if a silo is already present, €20,500–29,500 can be saved.

3.6.2.4 FINANCING

For the purpose of an initial cost–benefit analysis it is sufficient to assume that the annual costs of the investment are discounted at a single discount rate (generally over the same period as the technical lifetime of the installation). In practice, the financing of the investment will usually be arranged by means of a loan and partially with the buyer's equity. Also, the loan may have a different term from the technical lifetime of the installation. For the level of detail presented here (estimation) a full cashflow overview is not necessary.

The technical lifetime of a digester lies between 10 and 20 years; an average of 15 years is generally used. In order to calculate the annual costs of the investment, Table 3.9 can be used.

Table 3.9. Annual percentage of investment costs for various discount rates

Discount rate (%)	\multicolumn{11}{c}{Duration of investment (technical lifetime)}
	10	11	12	13	14	15	16	17	18	19	20
1	10.6	9.6	8.9	8.2	7.7	7.2	6.8	6.4	6.1	5.8	5.5
2	11.1	10.2	9.5	8.8	8.3	7.8	7.4	7.0	6.7	6.4	6.1
3	11.7	10.8	10.0	9.4	8.9	8.4	8.0	7.6	7.3	7.0	6.7
4	12.3	11.4	10.7	10.0	9.5	9.0	8.6	8.2	7.9	7.6	7.4
5	13.0	12.0	11.3	10.6	10.1	9.6	9.2	8.9	8.6	8.3	8.0
6	13.6	12.7	11.9	11.3	10.8	10.3	9.9	9.5	9.2	9.0	8.7
7	14.2	13.3	12.6	12.0	11.4	11.0	10.6	10.2	9.9	9.7	9.4
8	14.9	14.0	13.3	12.7	12.1	11.7	11.3	11.0	10.7	10.4	10.2
9	15.6	14.7	14.0	13.4	12.8	12.4	12.0	11.7	11.4	11.2	11.0
10	16.3	15.4	14.7	14.1	13.6	13.1	12.8	12.5	12.2	12.0	11.7

Example
An investment of €237,500 for an installation with a technical lifetime of 15 years will, at a discount rate of 5%, have annual costs of €237,500 × 9.6% = €22,800.

For possible subsidy schemes that can reduce these annual costs of the investment see Chapter 9.

3.6.2.5 OPERATION AND MAINTENANCE COSTS

The annual costs for maintenance are about 2–4% of the investment costs. This excludes the maintenance of the CHP unit, which can be arranged in a separate contract. The costs of this will be between €0.80 and €1.10 per operational hour (for CHP units < 100 kW). This includes the oil for the engine and overhaul, for a period of 10 years or for 60,000 operational hours.

The required labour for the operation of the digester is approximately 1/4 hour per day and an additional 1/4 hour per day for inserting the co-substrate. The costs or valuation of this type of labour will be in the range of €5–20 per hour.

3.6.2.6 INSURANCE AND TAXES

The annual costs for insurance (for circumstances beyond one's control) of the installation is about 0.5–0.65% of the total installation costs. Depending on your tax regime, an annual property tax may be applicable. The total costs for insurance and property tax will be in the order of magnitude of 0.5–1% of the total installation costs.

3.6.2.7 INTAKE AND END-USE OF CO-DIGESTED MANURE

The annual costs will include the costs for the intake of manure and/or co-substrates from an external source (including costs for transport to the installation). In some cases these costs may be negative (that is, an income). In addition to these intake costs, there are also costs resulting indirectly from this:

- Additional nutrients from external source. The additional nitrate, phosphate and potassium must be accounted for. If the buyer's own land does not have enough capacity to absorb these additional nutrients, supply to an external party is necessary. Costs for this will be in the order of €0.55/kg nitrate and €0.45/kg phosphate. These cost figures exclude the transport.
- Additional transport costs from the digester to the field/end-use. The extra mass that is added to the digester (including co-substrates from internal source) will have to be transported to the point of end-use. In the situation without digestion, there will only be transport costs for the manure, so all additional transport should be regarded as extra costs. The costs for transport will be in the order of €0–10/t.

Example
5000 t of cattle manure and 800 t of agricultural waste are digested annually. The cattle manure is from within the farm; the agricultural waste is from a neighbouring arable farm. All of the digestate can be used on the farm's own fields. The cost for intake is €1/t, including transport. The cost for the use of the extra nutrients is €0/t (on the farm's own field). The additional costs for transport (from digester to field) are €2/t. The net costs for intake and use of the co-substrate are 1 + 0 + 2 = €3/t.

- Savings on fertilizer use/better quality manure. The availability of nutrients in digested manure (mainly nitrates) has increased. Depending on the type of crop to which the (co-)digested manure is applied, the result will be a saving in fertilizer use. The saving due to the better quality of the manure is in the order of €0–1/t digested manure. In the future, digested manure may command a higher economic value in the manure market.

3.6.3 Benefits

In general an AD plant generates three types of revenue, profiting from the production and utilization of biogas and the improved fertilizing characteristics of the digested substrate. In some cases where specific waste streams are used as a co-substrate a fourth revenue component may result from waste removal fees. The major part of the income from operating an AD plant results from the value of the generated electricity. In most countries there are increased revenue tariffs for electricity from renewable energy sources that is fed into the grid. However, it is also possible to satisfy one's own electricity demand.

3.6.3.1 ELECTRICITY

For calculation of the benefits it is assumed that all the electricity is fed into the grid as sustainable electricity. This will be the case in most situations. For example, in Germany electricity from AD plants (<500 kW_{th}) is refunded at €99/MWh for a period of 20 years. In Austria the payments are even higher: €145/MWh in the same power capacity range, but for a period of only 13 years. In Spain electricity from AD plants is valued at approximately €60/MWh. Note that the benefits from electricity production may increase if the costs for the use of conventional electricity on the farm or company (for example at peak hours) are higher than the feed-in tariff. One may also choose to use a CHP unit with a bigger capacity in order to produce electricity only at peak hours. In this case larger biogas storage is also required.

3.6.3.2 HEAT

The residual heat can substitute for conventional production of heat. It is possible that not all of the residual heat produced by the CHP unit can be used, owing to variation in heat demand during the year (for example, there will be no heat demand on a hot summer's day) or variation during the day. Farms or companies with a sufficient heat demand will in most cases be able to use 50–60% of the residual heat. This largely depends on the type of farm or company.

The savings on conventional heat production will be in the order of €5–15/GJ, depending on the fuel used for heat supply and the costs of this. It is advisable to check the energy bill of the company or farm for the actual costs.

3.6.4 Cost-benefit calculation

Table 3.10 can be used to calculate the costs and benefits of an anaerobic digester. An example calculation is shown for the digestion of 5000 m^3 of cattle manure and 1000 m^3 (800 t) of agricultural waste per year, with an investment subsidy of 30%. The payback time in the example is 7.7 years. The payback time will be shorter if the amount of manure or co-substrates is larger, or if pig manure (with a higher biogas yield per tonne) is used.

Table 3.10.
Calculation of costs and benefits of an anaerobic digester. Example calculation shown for digestion of 5000 m³ cattle manure and 1000 m³ (800 t) agricultural waste per year

	Calculation method	Example	€/yr
Annual costs			
Annuity of investment costs	Total investment costs × (1 − subsidy %) × annuity %	€237,500 × (1 − 30%) × 9.6%	15,960
O&M: digester	Total installation costs × 3%	(€237,500 − €50,000) × 3%	5625
O&M: CHP	Operational h/yr × ~(0.8–1.1) €/hr	7500 h/yr × 0.9 €/hr	6750
Insurance and tax	Total installation costs × ~(0.5–1)%	€237,500 × 0.75%	1781
Labour	Total h/day × 365 × ~(5–20) €/hr	0.5 h/day × 365 days × 10 €/hr	1825
Net intake substrates	Sum of intake costs, transport and nutrient use	€3/t × 800 t	2400
Total annual costs			34,341
Annual benefits			
Electricity sales	CHP capacity (kW) × operational hours × feed-in tariff (€/kWh)	42.8 kW × 7500 h × €0.099/kWh	31,779
Heat use	Residual heat (GJ/yr) × usable % × ~€5–15/GJ	1295 GJ/yr × 50% × €10/GJ	6475
Savings on chemical fertilizer/better quality of manure	Amount of manure (t/yr) × savings (€/t manure)	5000 t/yr × €0.5/t	2500
Total annual benefits			40,754
Annual profit	Annual benefits − Annual costs	€40,754/yr − €35,091/yr	5663
Payback time	Total investment costs × (1 − subsidy %)/(Annual profit + Annuity of investment costs)	€237,500 × (1 − 0.3)/ (€5663/yr + €15,960/yr)	7.7

See section 3.6.2 for a breakdown of the investment costs and section 3.3.3.7 for the sizing of the components for this example.

3.6.5 Example of an anaerobic digester in practice

The German farmer Dietmar Epping has an operational digester on his farm. 'The main goal was to make a profit, but the environmental profits are also welcome,' says the enthusiastic farmer. He has only 55 dairy cows, but addition of corn makes the investment in the digester a profitable one. The operation of the digester does not require a lot of effort. 'Including the daily checks I need half an hour per day. The adding of the corn takes most of this time.' The farmer is satisfied with the operation of the digester. 'The installation has been running for half a year now without any problems, and has produced over 500,000 kWh. The payback time will be about five years.' The produced heat is used partly for heating his house: 'The cows don't need extra heating.'

3.7 References

1. T Al Saedi et al. Good practice in quality management of AD residues from biogas plants. IEA Bioenergy Task 24, 2001.
2. H J Bendixen. Hygienic safety: results of scientific investigations in Denmark, IEA Bioenergy Workshop, Stuttgart-Hohenheim 29–31 March 1999.
3. H Schulz. Biogas-Praxis. Grundlagen–Planung–Anlagenbau – Beispiele. Ökobuch Verlag, Staufen bei Freiburg, 2001.
4. R H C van der Leeden et al. Mestvergisting op boederijniveau. 's Hertogenbosch, 2003.
5. H. Neumann (ed.), Top agrar extra, Biogas – Strom aus Gülle und Biomasse. Landwirtschaftsverlag, Münster, 2002.

3.8 Further reading

Bundesverband der landw. Berufsgenossenschaften e.V. Hauptstelle für Sicherheit und Gesundheitsschutz: Sicherheitsregeln für landwirtschaftliche Biogasanlagen, 2002
Ecofys. Haal meer uit mest. Report, Utrecht, 2002.
T Fischer, K Andreas, K J Chae, S K Yim, K H Choi, W K Park and K C Eom. 'Farm-scale biogas plants'. Journal of the Korean Organic Waste Recycling Council, 2002, Vol. 9, No. 4, pp.136–144
U Marchaim. Biogas Processes for Sustainable Development. FAO, 1992.

4 Liquid biofuels

4.1 Introduction

> The use of vegetable oils for engine fuels may seem insignificant today.
> But such oils may become in the course of time as important as
> petroleum and these carbon-tar products of the present time.

These were the words written by Rudolf Diesel in the preface to his patent of 1912.

Whereas diesel engines and fossil fuels have been able to claim triumphant success all round the world, the use of liquid biofuels is today still at the start of its development phase, despite its advantages for the environment.

The use of these biofuels has penetrated the mobile engine market more than the stationary market. The main reason for this is the greater economic competitiveness – for tax reasons – of liquid biofuels.

Liquid biofuels are especially suitable for use in niche areas such as inland water transportation, or mobile and stationary applications in drinking water reserves, because of their compatibility with the environment and the low risk of endangering water bodies.

This chapter documents the status of technology found around the world for producing these fuels. However, the possibilities for further use of liquid biofuels are not forgotten: these are discussed in the chapters on the environment and market development. So far little operational experience has been gained in the area of liquid biofuels in stationary applications. Therefore this chapter will not go into the technical and operational details of project implementation, but will instead focus on background factors and general project mechanisms.

4.1.1 Who is this chapter aimed at?

This chapter is aimed at those who are interested in liquid biofuels and wish to inform themselves about the status of the technology for the production and utilization of liquid biofuels. Furthermore, best practice examples are presented to show that, assuming certain general conditions are met, the use of these types of fuel already makes sense, and can be economically attractive.

4.1.2 What information is provided in this chapter?

The first section provides an overview of the background and the many different environmental and socio-political advantages of liquid biofuels. It also takes a detailed look at the markets for the various biofuels in Europe, the USA and South America.

In the following sections the technical procedures for producing biofuels are examined and, in addition to discussing emissions properties, the market development perspectives are described. The possible technical applications for these fuels in transportation and stationary areas are also outlined.

4.2 Biofuels in transportation

Nearly 30% of carbon dioxide emissions from industrialized nations stem from the transportation sector. The movement of people and the transport of goods in the European Union lead to emissions of approximately 902 million tonnes of carbon dioxide into the atmosphere every year. These emissions are caused primarily by fossil fuels imported from other parts of the world (Figure 4.1).

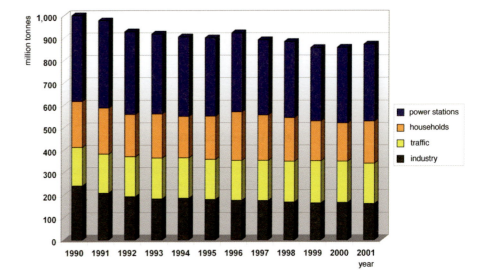

Figure 4.1.
Development of CO_2-emissions in Germany
Graphic: Dobelmann / www.sesolutions.de
Data: Energie Daten 2002

At present, the transportation and logistics sector still depends predominantly on a fossil fuel, namely petroleum. This observation, which may at first glance seem somewhat trivial, becomes all the more meaningful when one considers that the logistics for people and goods are at the heart of our economic systems. Supply shortages in this area, due for example to political developments, directly affect our economic cycles. Even small changes to fuel prices have consequences for the global economy's development.

The economy's dependence on fossil fuels is the easiest to understand in the fuel market. Fuel alternatives in this sector are scarcer than in other energy areas. The increasing globalization of world trade goes hand in hand with an increase in the transportation of goods and people. It is an accepted fact that fossil fuel resources are finite. Therefore we shall not be able to continue using these energy sources for ever. Moreover, the continuing mechanization of the developing and emerging countries has led to a leap in demand for fossil fuels. As the supply of these cannot be sustained, long-term economic problems are on the horizon.

These factors, and the often politically unstable and economically precarious situations in many petroleum-producing countries, are forcing many industrialized nations to look for alternatives, in order to be less dependent on petroleum imports.

To turn the tide, overall vehicle usage must be reduced, and climate-neutral, renewable fuels must be developed. This should be achieved in two ways. First, there must be self-imposed obligations to minimize consumption. For example, European automobile manufacturers are aiming for an average carbon dioxide emission level in all new cars of 140 g/km. Second, the use of alternative fuels made from renewable sources will break this dependence on petroleum-producing countries and place more emphasis on domestic resources.

4.2.1 The market for liquid biofuels

The European Union (EU) is the worldwide leader in the production of biofuels. A total of 2,100,000 tonnes of biofuels are produced and used in the transportation sector annually. The main producers in the EU are Germany, Italy, Austria and France, followed by Spain and Belgium.

The European Commission wants a fifth of the transport of goods and people in the EU to be based on renewable biofuels by 2020. Thus 20% of mobility in the EU would be catered for in a climate-neutral way.

Significant differences are apparent in the EU member states regarding the market penetration of biofuels. This is essentially due to varying strategies for technical realization and tax incentives. Regarding technical realization there are two options: using a blend of biofuels with fossil fuels, or using pure biofuels. Yet the key to the development of the European biofuels market is the cost of these fuels.

Currently, one of the main distinguishing features of biofuels is that they have higher raw material costs than fossil fuels. At present many of these raw materials find their primary sales markets in the foodstuffs and cosmetic sectors. Vegetable oil producers, for example, are constantly confronted with the decision of whether to put their product on the foodstuffs market or the fuel market.

The production costs of these fuels range from €0.02/MJ to €0.05/MJ, and are therefore significantly higher than the production and distribution costs for fossil petrol and diesel. The difference in costs between renewable and fossil sources is significantly greater in the fuel market than in the electricity and heat markets. Accordingly, biofuels need to be subsidized in order to be able to compete in the market.

A logical alternative, also in line with environment policy, would be to exempt these fuels from the mineral oil tax levied on petroleum. This would set off the cost disadvantage that currently exists in production, and which has to be passed on to consumers. By levelling out the costs, or even slightly favouring biofuels, sustainable demand could be stimulated in the market, and the environmental benefits could really come to fruition.

4.2.2 The advantages of biofuels

The cultivation and processing of liquid biofuels emits less climate-relevant carbon dioxide than that of fuels from fossil sources. When looking at energy sources overall, as well as at the individual dangers for water, the climate and human health, liquid biofuels compare very favourably with fossil fuels.

Biofuels are inherently more biodegradable than fossil fuels, and therefore represent a much lower threat for inland and coastal waters than fossil fuels such as petroleum. This, and the fact that biofuels are mostly produced in the same region as they are consumed, means that the risk of danger resulting from transportation is greatly minimized.

Yet the advantages of biofuels are not just limited to the environment. The presence of biofuels on the fuels market also brings substantial socio-economic advantages. Biofuels are an important feature of overall plans for rural development in Europe.

Figure 4.2.
Water body protection by Biofuels
Photo: UfOP / www.ufop.de

In this regard the long-term labour potential of these fuels is a major factor. The manufacture of biofuels and raw materials can pave the way for multi-functional farming, which can create new sources of income and jobs in the rural areas. For example, assuming the EU had a sustained demand for 7 million tonnes of biofuels, 2000 jobs would be created in plant cultivation itself, and 7000 jobs would be generated in processing. Overall, biofuels could be the source of up to 120,000 new long-term jobs.

4.2.3 Areas of application

Although occasionally used in combined heat and power or pure heat generation, liquid biofuels are primarily used in the fuel sector. This is partly because liquid fuels, owing to their physical properties such as pumpability and high energy density, have generally proved their worth in the transport and logistics sector. It is also partly because the economic hurdles to stationary use are still, on average, too high.

Fuels from biomass offer an interesting alternative to fossil mineral oils. Even today they provide the means to make climate-neutral fuels for all purposes. In addition to pure hydrogen, a range of carbon-based liquids can be considered for renewable fuels, including natural or secondary/used vegetable oils and their esters (biodiesels) and the group of alcohols such as methanol and ethanol, as well as other hydrocarbons such as synthetic petrol/diesel fuels. The renewable fuels already produced from biomass in larger quantities for the transportation sector are vegetable oil, methyl esters and ethanol.

However, the production of liquid biofuels is in its infancy, and the development possibilities are substantial. Technical scenarios forecast that 25% of the EU fuel market can be modified to suit renewable fuels. It would already be possible to reach this target with existing technologies if there was a moderate reduction in the average fleet consumption, and if 50% of the biomass available was used for fuel production.

If the discussion is extended to synthetic biomass fuels, which are still in the research stage, the renewable share of the market could be as much as 45%. The EU would then achieve significant reductions in its carbon dioxide emissions with regard to its climate protection obligations. Furthermore, a considerable degree of independence from petroleum could be achieved.

In the transportation sector two types of engine have prevailed for powering automobiles: the spark ignition engine, fuelled with petrol, and the auto-ignition diesel engine, fuelled with diesel. Biofuels can be produced for both engine types.

Both natural oils and fatty acid methyl esters can be used to fuel diesel engines. Today the fuels most common in the market are fatty acid methyl esters, because they can be used in traditional diesel engines without the need for complicated technical modifications.

Fatty acid methyl esters are produced on the basis of vegetable oils. These vegetable and animal oils and fats are subjected to the esterification process using methanol. In Germany considerable market success has been achieved with rapeseed methyl esters, the so-called biodiesels.

Yet to achieve maximum market penetration, biofuels must be made available for cars with petrol engines as well. Alcohols such as methanol and ethanol are suitable for these gasoline engines. Next to the common petroleum-based production these can also be produced from anaerobic digestion of biomass.

Technically speaking, biofuels such as biodiesel and bioethanol/methanol could replace mineral oil-based fuels entirely. When assessing the overall ecological benefit, it is important to take into account the amount of energy needed to produce one unit of fuel. In this regard both renewable fuel types do very well: see Figure 4.3.

Another way to fuel cars, without the combustion engine, is to use energy from electricity. This can be provided from energy storage (batteries) or from energy converters (fuel cells). The advantage of electricity-fuelled cars is that they are

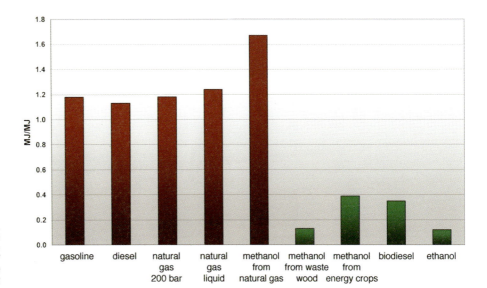

Figure 4.3.
Specific energy consumption in fuel production
Graphic: Dobelmann / www.sesolutions.de
Data: Ludwig Boelkow Systemtechnik

emissions-free when driven. They do not harm the ambient air with exhaust fumes during operation. However, contrary to claims sometimes made by the industry, they are certainly not completely emissions-free. The energy required for charging or fuel storage usually results in carbon dioxide emissions. Yet there are exceptions – for example when the electricity needed is produced directly from renewable energy sources such as the sun, the wind or hydropower.

The requirements of a modern fuel are numerous and diverse. The most important are as follows:

- The production costs must be acceptable.
- The fuel must be capable of being produced in sufficient quantities.
- The infrastructure for transportation and distribution must be financially viable.
- The fuel must be suitable for combustion engines, fuel cells and other energy converters if needed.
- It must have a significant potential to reduce CO_2 emissions.
- It must have a low overall emissions potential.

Figure 4.4.
European oil seeds
Photo: UfOP / www.ufop.de

*Figure 4.5.
Rapeseed harvest
Photo: UfOP / www.ufop.de*

The processes for manufacturing biofuels are technically advanced. They are already being widely utilized in the chemical industry and in the production of foodstuffs. The technical processes used are generally so advanced that an increase in the demand for biofuels will not trigger any large-scale reductions in the process-related and technical production costs.

Thus the cost structure of biofuels is determined less by the production processes and more by other factors such as the price of raw materials or secondary raw materials. In the area of natural vegetable oils, usage as a fuel competes noticeably with the production of foodstuffs and animal feed. There are other possibilities for recyclable materials – use in biogas plants for example – so there is no potential for cost benefits in this area either.

4.3 Process for producing liquid biofuels from biomass

*Table 4.1.
Raw materials used to make liquid biofuels*

Liquid biofuel	Feedstock	Technique	Application
Biodiesel	Plant oils: Oilseed rape Sunflower oil Colza oil Waste oil (cooking oil)	Pressure extraction Etherification with methanol	In pure form or blended with conventional diesel
Bioethanol	Sugar beets Cereals Other crops Plant waste products Wood, straw	Alcoholic fermentation	Component in gasoline or pure as motor fuel
ETBE	Bioethanol (derivative)	Reaction with isobutylene over catalyst	Blending component in gasoline up to 15%
Biomethanol	Lignocellulosic materials Biodegradable fraction of waste	Thermo chemical process	Equivalent to fossil methanol
MTBE	Biomethanol (derivative)	Reaction with isobutylene under catalysis	Blended as component

Table 4.1 lists the most important raw materials used to make liquid biofuels. The first four biofuels listed in this table are currently available on the market. Methanol, MTBE and other synthetic fuels are still in the trial and development phase. Other

natural or waste vegetable oils are also used in many scientific studies, but will be considered only briefly here, as they are not yet market relevant.

In the future, the biofuel industry is expected to focus much more on secondary raw (recycled) materials markets as a cheaper source of raw materials. The use of waste as a source of secondary raw materials, and thus as an alternative to waste disposal, can create a cost-effective basis for fuels, but this possibility always depends on the individual waste disposal costs. Furthermore, there is no development potential in this sector in the medium term, as a complete distribution of the secondary raw materials can be expected fairly soon.

4.3.1 Natural vegetable oil

Some types of crop, such as rapeseed, sunflowers and olive trees, have a high content of vegetable fat, which can be used in technical processes. Sunflowers and rapeseeds are reaped with a combine harvester, which separates the oil seeds from the rest of the plant.

Two technical processes are available for producing vegetable oils from all the crops mentioned. Which one is used depends on the size of the production plant.

In small and decentralized oil mills the oil is extracted using mechanical auger machines. These mills process between 0.5 and 25 tonnes of rapeseed or sunflower seeds per day. Up to about 80% is extracted from the plants. The cost of purchasing oil mills of this daily capacity is between €8000 and €15,000.

In industrialized production an additional extraction procedure is carried out on the plant residue, following mechanical extraction. Solvents are used to release the oil from the crushed residue. Then the solvents and oil are separated. This procedure increases the yield of usable oil even more, although the oil has to go through intensive purification after this treatment. Industrial oil mills with this sort of process management have an oil yield of up to 99%. They process up to 4000 tonnes of oil seeds per day.

Following extraction there is an intensive purification process of the natural vegetable oil in both types of production facility. The solids remaining in the oil after pressing – up to 6% – are separated. In small plants this can be achieved through continuous sedimentation. Larger industrial facilities generally use centrifuges for this.

In northern Europe rapeseed dominates the production of vegetable oils for the fuel and foodstuffs markets. It delivers an oil yield of about 1150 litres per hectare of farmland. Rapeseed can only be sown on the same piece of farmland every four years. In southern Europe the sunflower is the biggest supplier of vegetable oil. Both types can be processed into fuels without restrictions.

The remains from the rapeseed meal are used as natural animal feed for cattle breeding. The sale of this residue as animal feed is essential for improving the economics of producing rapeseed oil. About 1900 kg of rapeseed meal per hectare of rapeseed field can be passed on to producers of animal feed.

4.3.1.1 TECHNICAL PROPERTIES AS A FUEL

At 77–78% of its weight, natural vegetable oil has a very high carbon content. Hydrogen (12%) and oxygen (10%) make up the rest. Vegetable oils are natural products, and are therefore readily biodegradable. This property can sometimes be an obstacle to their use as a fuel, because oxidation and polymerization processes can start in the storage tanks. These change the oils' fuel properties, with the result that, even under favourable storage conditions, the maximum storage life of natural vegetable oils remains limited to 6–12 months.

Vegetable oils react differently to cold conditions than do refined fuels. In decreasing temperatures they become more and more viscous, to the point where they can solidify.

In countries with cold winters the considerable effects of frost need to be considered. It means that vegetable oils can either only be used for driving when blended with traditional fuels, or a maintenance of the fuel temperature above 5–10°C must be guaranteed. This means keeping the oil tank at a temperature that guarantees the right viscosity for the fuel injection system. These factors make driving with natural vegetable oils technically more complicated than driving with conventional fuels.

4.3.2 Biodiesel

Natural vegetable oils cannot be used in conventional diesel engines without modifications having to be carried out. Therefore if the acquired vegetable oil is to be brought to the traditional fuel market without the need for engine modifications, it must undergo the esterification process to convert it to biodiesel.

Figure 4.6.
Biodiesel at the gas station
Photo: UfOP / www.ufop.de

Vegetable oil, once it has been through the esterification process, is called biodiesel. In this process the fat molecules are separated into three individual fatty acid ester chains. The physical properties of the vegetable oils are changed in such a way that they correspond to those of conventional diesel fuels. After esterification the molecules have a viscosity that is similar to that of a normal fossil diesel fuel.

Biodiesel can also be produced from used cooking fats and oils as well as from natural vegetable oils. These fats must be collected centrally, and have to undergo special purification treatment before esterification can deliver the desired quality.

The esterification process requires alcohols, in most cases methanol. Glycerine emerges as a by-product from esterification and, after purification, can be used in the chemical industry as a base material. The processes used for the production of these fuels are well established in the foodstuffs industry, and have been optimized for the fuel market in the last few years. This makes it possible to adhere to the strict quality requirements of modern fuels and to achieve uniform quality.

4.3.2.1 TECHNICAL PROPERTIES AS A FUEL

In most cases the use of biodiesel does not require any adjustments to the engine. Moreover, for the most part, it is possible to use a fuel blend and the existing injection system. However, technical alterations can become necessary when the rubber seals are not made of fluorocarbon rubber. Although in its viscosity the biodiesel produced has almost identical properties to those of fossil diesel fuel, fatty acid methyl esters behave differently from chemical polymers.

Biodiesels attack and dissolve some synthetics used in cars, including fuel pipes and seals. Any components made of nitrile rubber are affected. In cases of prolonged contact with biodiesels they can swell and macerate. If manufacturers use fluorocarbon rubber components in the first place, or if they are installed during a refit, there are no such problems. Fluorocarbon rubber is resistant to attack from biodiesel, and is just as resistant to fossil diesel fuel. In this way the manufacturer allows the customer the possibility of driving with diesel or biodiesel without having to make any modifications.

4.3.3 Ethanol

Ethanol can be produced by fermenting sugary crops such as sugar cane, millet and sugar beets, and starchy vegetables such as maize, corn and potatoes. Sometimes lignocellulosic biomass, containing celluloses, lignin and hemicelluloses, is also used for producing ethanol: examples are wood waste, straw and plant waste.

Sugary plants can be mashed straight away and fermented to ethanol with alcohol fermentation. Starchy plants such as corn must first go through enzymatic saccharification, before alcohol fermentation can be carried out. Whereas ethanol production in the 1980s concentrated on using starchy and sugary plants, research and development practices today focus on lignocellulosic biomass. This is often more cost-effective, because lignocellulosic waste is available on the market and is not used in the foodstuffs sector.

Production from non-sugar and non-starch plants could mean a breakthrough for the production of ethanol from biomass.

Lignocellulosic biomass can be used for ethanol production only when broken down into glucose beforehand. During the transformation from lignocellulosic biomass, decomposition using steam enlarges the surface area, thereby creating the right conditions for producing the sugar compound via micro-organisms.

Hydrolysis is the next step. The lignocelluloses are transferred to a compound of glucose and other types of sugar. Glucose materials are then fermented into alcohol using yeast, in airtight conditions. It is also important, when using other bacteria and fungi, to ensure the right living conditions for the organisms. Parameters such as temperature and pH values are crucial for the success of the fermentation.

The alcohol-containing compound produced by fermentation is called mash. Ethanol is isolated from it by several stages of distillation (rectification). With multiple-step distillation the necessary fortification of the ethanol to a purity level of 96% can be achieved. The remaining wash consists of water and organic materials.

4.3.3.1 TECHNICAL PROPERTIES AS A FUEL

The usually volatile ethanol is very suitable as a fuel for direct-injection gasoline engines. Owing to its high octane number, from 110 to 130, it is a very knockproof fuel. As a result, engines operated with ethanol and optimized for this type of fuel can use a very high level of compression.

Distilled ethanol can also be used as a blending component for fuelling standard gasoline engines. The properties of ethanol increase engine efficiency and reduce fuel consumption. Moreover, as the resulting octane number is higher, an engine-friendlier fuel is produced. Up to 10% volume can be added without technical modifications having to be carried out to the complete fleet of cars. Consequently, the blend most often available on the market is a 10% volume ethanol and 90% volume petrol compound. Higher concentrations of ethanol – up to 85% – and indeed pure ethanol are also a suitable fuel source for direct-injection gasoline engines, although this requires modifications to be made to the common standard of engine.

4.3.4 Fuels from synthesis gas

In addition to extraction and fermentation procedures, thermochemical approaches in particular have the potential to generate fuel via the production of synthesis gas. However, the processes are in various stages of research at the moment, so synthetic fuels are not expected to be launched on the market for a few years yet.

The advantage of this type of fuel is that it can be produced from a relatively wide spectrum of raw materials. This diversity allows a production plant to be economically effective despite fluctuating raw material prices.

Another possibility for these synthetic fuels is the production of a tar-free synthesis gas, which can also pave the way for a fuel economy based on hydrogen. Synthesis gases, with or without carbon monoxide elements, offer the greatest possible flexibility, not only in terms of the raw materials used, but also in terms of the products produced. Many of the fuels already used and many of the fuels planned for the future can be produced from a pool of raw materials based on synthesis gases.

Methanol, methane, hydrogen and synthetic petrol and diesel fuels can all be produced via the synthesis gas intermediate stage. Synthetic fuels in particular have the potential to display more environmentally friendly emission behaviour as long as combustion properties are identical or better. Research is being carried out in the automobile industry in many places into developing procedures for producing synthetic fuels alongside the hydrogen option.

A great deal of research is still needed into the production of synthesis gas from the sub-stoichiometric oxidation of biomass. The procedures currently used for producing gas from biomass were not designed to produce synthesis-capable gases, but to convert the gases produced into electricity. The technology for the gasification of carbon, for example, is state of the art. Unfortunately its direct transfer to biomass gasification in smaller decentralized plants (< 50 MW) has not seen immediate success.

Many questions regarding the production of synthetic fuels have not yet been conclusively resolved. They include the quality and stability of the fuels produced, the production costs, the possible yield of the raw materials, and the project's overall energy efficiency.

4.3.4.1 PRODUCING SYNTHESIS GAS

The initial step in producing gas from biomass is the thermochemical conversion of the base fuels. Air, oxygen, steam and hydrogen as well as mixtures of these gas components can be used as technical gasification mediums for this process. The fuel production that follows is a catalytic process, so the synthesis gas must satisfy the particular requirements and should show a high level of purity.

Research is currently being carried out into solutions and strategies for the gasification of biomass. Hitherto, the production of a synthesis-suitable gas from biomass, which is free from tar, dust and inert gas and yet hydrogen-rich, could not, with enough stability for the process, be proved successful in plants of < 50 MW. The use of gas conditioning units allows such qualities, but in a fuel market competing on price these are rejected on cost grounds.

Economically, small plants are crucial for the use of gasification technology for biomass, for logistical reasons. These must be designed in such a way that the use of pure oxygen is not needed. Supplying gasification facilities with pure oxygen or building an air fractionation facility is not suited to decentralized biomass gasification for reasons of costs.

Another essential requirement of the synthesis gases used to produce fuel is the hydrogen content. Sometimes it is well under 50%, especially with autothermic gasification procedures where the energy is generated from the processed fuel. Allothermic gasification procedures, which use heat supplied externally, have a far better potential of being used to produce synthesis gases, as they promise a higher quality of gas.

In the allothermic process, external heat is transferred to the reactor. This means it is not necessary to provide oxygen as a synthesis gas, and the resulting product gas has a usable hydrogen content. However, this procedure does require considerable quantities of external energy as, in contrast to autothermic procedures, it does not fulfil its energy requirements itself.

4.3.4.2 CURRENT STATE OF RESEARCH
Germany has a test plant for the gasification of biomass, which could be a fundamental element in the base material for producing synthetic fuels. In the first step of the process, dry biomass such as wood, straw or organic sludge is turned into a carbon-monoxide-rich synthesis gas by means of sub-stoichiometric gasification. In the second step synthetic fuels can be produced from this synthesis gas once it has been purified to remove particulate contaminants.

Hydrocarbons can also be produced from synthesis gas using Fischer–Tropsch synthesis. In reaction conditions of between 220°C and 240°C, and with a pressure of about 25 bar, long-chain hydrocarbons can be converted into short-chain ones by using iron- and cobalt-based catalysts. Hydrogen must be made available for this process. However, in decentralized fuel production this process is technically very complicated, because a wide range of products emerges during Fischer–Tropsch synthesis.

4.3.5 Methanol
Methanol is used in a wide range of areas, from blending with conventional fuels (without changing the technology used) to being used pure as a fuel. It can be used in traditional combustion engines (Figure 4.7) and in direct methanol fuel cells, but it can also be used as a base product for making biodiesel from vegetable oils.

Figure 4.7.
Methanol at the gas station
Photo: DaimerChrysler /
www.daimlerchrysler.de

Methanol can be made from biomass, but this is still in the trial stage. Important for methanol production is the conversion of solid biomass into a gaseous form. This can be done via the production of synthesis gas as described in the previous section, or by using biogas from landfills or biogas facilities.

The production of methanol is a cost-intensive chemical process. Therefore in current conditions only waste biomass such as old wood or biowaste is used to produce methanol.

4.3.5.1 PROPERTIES
The chemical formula of methanol is CH_3OH. Methanol has a lower volumetric heating value than conventional fuels such as petrol and diesel.

There is one decisive drawback to the use of pure methanol over conventional fuels. Although methanol is liquid, it has corrosive properties. Therefore it cannot be distributed via the existing distribution network of petrol stations designed for petrol and diesel.

In the prevailing environmental conditions and atmospheric pressure, methanol is liquid in its chemical structure. This compact structure gives methanol cars with the same size of tank longer ranges than cars that operate on natural gas or hydrogen. However, methanol is a powerful environmental poison and dangerous to human beings, because it can be absorbed even through by skin contact. The fact that the fuel is highly soluble in water is also a significant cause for concern as far as the environment is concerned. Once methanol escapes into the ecosystem it is immediately dissolved by water and cannot be retrieved easily. However, in terms of biological degradability, it fares significantly better than petrol.

Energy efficiency for producing methanol from biomass reaches values of up to 55%. Alternatively, if used at lower degrees of efficiency (for methanol), the synthesis gas that remains can be used to generate electricity.

4.3.6 Hydrogen from biomass

In many energy scenarios of the future hydrogen is considered an important source of energy. However, hydrogen does not come on its own. It is combined with other elements, such as oxygen to make water or carbon to give methane, and must first of all be separated from these elements using energy. This means that hydrogen is only as ecological as the energy sources used to produce it.

Hydrogen is an odourless gas, which has a density of about 0.09 kg/m^3 and is significantly lighter than air. Its boiling point is –253°C. This makes it technically considerably complicated to store, transport and distribute. Furthermore, hydrogen has a low volumetric energy density, which means that cars with a range of over 500 km need liquid stocks. Storage in gas-pressurized containers is still possible for short-distance ranges.

When hydrogen is used in combined heat and power engines (CHP) or fuel cells, the only harmful substances that can emerge are nitrogen oxides, if oxidation with air occurs. So hydrogen is an appealing source of energy for environmentally sensitive areas, such as city transport.

4.3.6.1 PRODUCING HYDROGEN FROM BIOMASS

Although hydrogen is essentially acquired through electrolysis, it is also possible to produce it by using biomass. By means of partial oxidation, a combustible gas can be produced that consists largely of carbon monoxide and hydrogen. The synthesis gas produced is also called weak gas because of its low heating value. Should there be the intention to produce pure hydrogen suitable for fuel cells from this gas, a complicated purification process is necessary. Dust and tar as well as any carbon monoxide must be removed.

4.3.6.2 CURRENT STATE OF RESEARCH

The gasification of biomass as a source of hydrogen has been a topic of research for a long time. The technical problems related to the procedures used concern mainly the complete gasification of the different types of biomass and the subsequent gas purification. The production of hydrogen from biomass is not yet market ready.

4.4 Costs of liquid biofuels

Renewable fuels could replace fossil fuels, and would make a sustainable contribution to environmental and climate protection. However, to penetrate the market, renewable fuels would have to adjust to, and be subject to, the same economic conditions and realities.

At both the European and national levels some tax policies and support programmes exist to compensate for some of the added costs of these fuels – tax concessions for example. However, these instruments compensate only in part. The main criterion for competing successfully in the market is still low production costs.

The current costs of fuels based on biomass, at around €0.8/litre, are significantly higher than the production and distribution costs of petrol and diesel. Whereas the cost

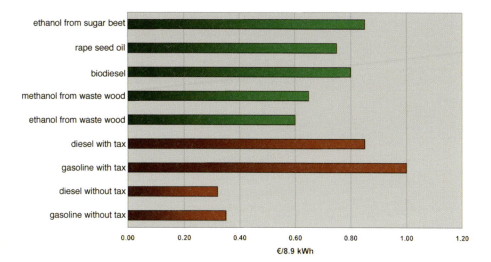

Figure 4.8.
Average fuel costs
Graphic: Dobelmann /
www.sesolutions.de
Data: Ludwig Boelkow Systemtechnik

of electricity from renewable sources is already acceptable for end-users, the difference in costs between renewable and fossil fuels is significantly greater (Figure 4.8).

If, in an analysis of whether to use renewable energies – from biomass in particular – to produce heat, electricity or fuel, only the costs of avoiding CO_2 emissions were considered, the substitution of fossil energy with biofuels to create heat and electricity would appear considerably more advantageous than the substitution of crude oil with biofuels to produce fuel. However, if the catalogue of criteria is widened to include diversification and supply security, then the result of the analysis changes in favour of fuel production from renewable energies.

When evaluating the use of renewable primary energy sources in transportation, future price developments are just as relevant as current ones. The relations between price movements in the two sectors, mobile and stationary, are important, too. When demand for individual products is stimulated by increasing compensation rates, the areas of use can shift between the CHP production of electrical energy and the fuel market.

If political developments or a scarcity of resources were to lead to a premature and over-proportionate rise in crude oil costs, the day when renewable fuels become totally economically efficient could be delayed, even without political and tax incentives. However, economic efficiency is just one of the many aspects of renewable fuels to be considered. The benefits of these fuels, such as the better environmental compatibility and the minimizing of health dangers, mean that a faster and more comprehensive market launch is already possible in some areas.

4.5 Liquid biofuels market development

Every price increase for fossil fuels paves the way for another market launch of liquid biofuels. It is easier for many biofuels to become competitive on the heavily taxed fuel market. In the stationary market, which is also supplied with solid biofuels, this is considerably harder for liquid biofuels. Consequently the development of market potential will focus primarily on the fuel market.

The following paragraphs describe current market development trends for the most important biofuels.

4.5.1 Natural vegetable oil

As more successes in the technical trials are recorded, the use of natural vegetable oil increases. At the moment most of the engines are still on trial runs, or are in the first stages of serial use. The further market development of this type of fuel will depend heavily on the successes of these projects.

In contrast to fossil fuels and biodiesel already developed for the mass market, natural vegetable oil will feature more strongly in regional markets. This is because of the technical properties mentioned in section 4.3.1 (that is, the limited storage time), and also because – in order to guarantee the economic efficiency over a longer period of time – long-term supply contracts for bioenergy projects have to be secured.

4.5.2 Biodiesel

In principle there are two different approaches that can be taken to secure the wide use of biodiesel in the national fuel market: whereas German law prefers a biodiesel to be used in the pure form, in France biodiesel blended with fossil fuels carries the tax advantages.

Alongside the general need to make the conscious decision to use biogenous fuels at filling stations, these strategies also have ecological considerations. The use of pure biodiesel in conjunction with oxidation catalysts can lead to reduced exhaust fumes emissions. It can create optimal conditions for niche applications such as transport or water protection areas.

Logistically speaking, the addition of biodiesel to fossil fuel is simpler and incurs hardly any additional costs in the production and distribution of the fuel. However, the benefits in terms of emissions are negligible.

The ecological advantages of using biodiesel should be given more attention: for example, use in inland water transportation on sensitive lakes, in water protection areas and for city transport is particularly recommended.

The use of biodiesel for CHP energy production in stationary applications is being held back mainly by economic obstacles. Because of the raw materials and production costs, biodiesel is considerably more expensive than the tax-favoured heating oil. Support programmes could break down the barriers to economic efficiency in this regard.

4.5.3 Ethanol

In most countries in the world the production of fuel alcohol still plays a secondary role. In France, Brazil and the USA alone there are state programmes sponsoring the use of ethanol. Yet it is only now, in the context of increased oil prices and newly aroused interest in an environmentally friendly energy supply, that the subject has been receiving more attention.

The role of ethanol in the fuel market can be found in the area of fuel additives. This has to do with economic efficiency. As a rule, higher market prices can be attained when it is sold as a fuel additive than when sold as a pure fuel. Furthermore, for the quantities used commercially as an additive, no technical modifications to the cars are needed.

Even if hitherto Europe has not clocked up any great successes in the use of ethanol, internationally there have been considerable signs of success. Brazil, for example, has been running a successful ethanol-fuel programme since 1975. Thanks to the Pro-alcohol Programme, 13.5 billion litres are currently being pumped into the fuel market annually. This makes Brazil the leader of fuel-alcohol-producing countries; it produces 42% of its fuel market with renewable sources.

In Brazil the production of alcohol for the fuel market is based solely on sugar cane, which creates 40% of farming revenue. Ethanol production from sugar cane was promoted as part of the Pro-alcohol Programme, as was the development of pure ethanol engines, which need only 4% additives. The sinking prices on the world market for crude oil experienced in the 1980s had a negative impact on the programme, with the result that, to shore up the alcohol market, it was made compulsory to add 24% ethanol to petrol supplied from petrol pumps.

The USA also has an ethanol programme for road traffic. In 2001, 6 billion litres of alcohol were pumped into the fuel market and replaced 1.5% of petrol sales in the USA. The strategy in the United States is to insist that a maximum of 10% ethanol be added, and that this comes predominantly from maize. The current programme is showing an upward trend due to increased environmental awareness. The background to these activities is the strategy to reduce the burden on the environment caused by carbon monoxide (CO) and ozone (O_3) by adding oxygen-rich fuel components such as ethanol or its derivatives.

The use of ethanol in stationary power plants is technically possible. Because it is used mainly as a fuel additive, it has not so far made its mark as a pure product on the existing ethanol markets. This is especially true for use in stationary combined heat and power.

4.6 Using liquid biofuels for mobile applications

4.6.1 Natural vegetable oil

At present, knowledge of the use of natural vegetable oils in modern engines is limited. Important data records on the long-term conduct of these fuels, together with proof of lifespan on test benches and in car use under various conditions, are not available.

Because of this unsatisfactory situation, in Germany round-robin tests with 110 tractors from various producers with a performance capacity between 50 kW and 150 kW were carried out. This led to the finding that natural vegetable oils can be used as a fuel in adapted diesel engines.

For natural vegetable oils to be used in diesel engines, technical modifications must be made. In addition to preheating the fuel, the most important requirement is to adapt the fuel injection system to the viscosity of natural vegetable oils. Additionally, the fact that fuel distribution behaviour is different usually makes a different pistons set-up necessary.

In terms of the most essential engine components, the diesel engines used with vegetable oil as a fuel are identical to those used with fossil fuels. They use direct injection systems such as nozzles as well as indirect injection processes such as antechambers and swirl chambers to bring the fuel to the combustion chamber.

The antechambers and swirl chamber methods are generally well suited to viscous fuels such as vegetable oils because of their two-tier turbulent combustion. When using the direct injection process, the pistons have to be adapted by adding a hemispherical fuel basin, into which the vegetable oil can be injected tangentially.

At low temperatures, because of its high viscosity, it is difficult to process vegetable oil into a compound that can be ignited in diesel engines. To avoid ignition problems in cold engines, conventional diesel fuel is used for the ignition process, and the natural vegetable oil is only injected following a warming-up period for the engine. This means that a two-tank system must be in place.

Another problem with this natural viscosity of vegetable oil is that, in a fuel injection system, it cannot be easily atomised into an ignitable compound. Even if this is relatively unimportant for short runs in normal conditions, in the long term it can lead to unequal ignition and therefore more accumulation of debris on cylinder liners, pistons, valves and injection nozzles, the latter of which are already under particular strain.

4.6.2 Biodiesel

The use of biodiesel in volume-production diesel engines is generally possible. However, it is important that the biodiesel be approved by the engine manufacturer. This approval can either be issued ex works or once a conversion kit – essentially replacing rubber seals and other parts made of nitrile rubber that come into contact with the fuel with ones made of fluorocarbon rubber – has been used.

When using biodiesel in practice, the following should be noted to help guarantee smooth, long-term operation:

- If the car changes to biodiesel after a long period of time during which only petroleum diesel has been used, a change of the fuel filter may be necessary. As biodiesel behaves like a solvent, diesel fuel residua can be dissolved, which can lead to filter blockages.
- For the same reason, lacquered areas that come into contact with the biodiesel should – as is the case for conventional diesel – also be wiped down immediately.
- If biodiesel is used in non-approved cars, some rubber or synthetic materials can be damaged in certain circumstances and over a longer period of use. For example, fuel tubes can macerate. Tubes made of fluorocarbon rubber, already used as standard in approved cars, can redress this. Information about the types of material used can be supplied by the appropriate garage. Regular checks of the fuel system and, if needed, changing of the materials affected can be carried out quickly and affordably.

- Intervals between oil changes should be carried out according to the manufacturer's guidelines. A thinning of the engine oil by the fuel can occur in commercial vehicle engines. However, this generally happens only when the engine is driven over a longer period of time and under little strain.

In Germany DIN 51606 is the basis for quality assurance for fuels that have undergone esterification. To fulfil warranty claims, and for operation in general, all diesel engine manufacturers that issue approvals for these fuels demand fuels that satisfy this standard to the full.

4.6.3 Ethanol

Ethanol is seldom used in Europe and the USA as a pure fuel; only in Brazil is this method used. The reason for this is the fuel's high evaporation rate, as this reduces the capacity for cold starts.

The problem of cold starts at low temperatures is not very relevant in warm climate zones. In colder climate zones such as northern Europe the cars sometimes have to be equipped with an extra petrol tank, which takes over cold starts.

The use of ethanol when blended with conventional petrol does not require any technical modifications to the engine. Its technical qualities increase the fuel's octane number by two numbers per 10% proportion of weight. Ethanol–petrol-mix fuels are generally engine-friendlier than pure petrol fuels.

4.7 Using liquid biofuels for stationary applications

4.7.1 The basics

The liquid biofuel with the greatest potential in stationary combined heat and power is natural vegetable oil. This can be produced in agricultural cooperatives using simple means and prepared for use in adapted engines. Therefore these guidelines will focus mainly on natural vegetable oils. However, the general statements in these guidelines would also apply to a bioenergy project being carried out with other fuels.

Generally speaking, all liquid biofuels can be used in stationary applications where weight – a factor always relevant to mobile application – can be disregarded. This opens the door to some technical possibilities where many of the environmental advantages these fuels offer really begin to take effect. Examples of this are the double-layer tanks, which exclude the possibility of fuel creeping out and endangering water, or the fitting of soot filters and other waste gas treatment systems.

Combined heat and power is considered an important part of environmental protection all over the world because of its high level of energy efficiency. This also applies when fossil fuels are used in the long term instead of renewable fuels. This fact, combined with the economic efficiency of the plant operation essential for CHP projects, makes the use of liquid biofuels for stationary application more difficult.

The design of stationary vegetable oil engines is the same as that used for mobile applications. The advantage of combined heat and power for stationary engines is that many of the technical problems associated with vegetable oil, such as cold starts, can be eliminated. Moreover, in combined heat and power, enough heat is usually available to acquire sufficient control of the viscosity problem observed in low temperatures. The engine units in CHP plants are mostly already in a warm or preheated state when they start.

The CHP engines are designed to work with optimum performance for as many hours per year as possible, beyond 4000 operating hours, in order to ensure a high level of efficiency. This makes optimal and regular maintenance intervals possible, when expendable parts can be replaced.

Combined heat and power with vegetable oil is the same as traditional CHP units in natural gas or fuel oil operations. A combustion engine is coupled with an electric generator, which changes the engine's mechanical energy into electricity. The electricity energy efficiency of the primary energy used in this procedure is about 30%. The heat being generated in the exhaust fumes or in cooling system processes of the engine can then be uncoupled via heat-exchanging devices and made available to buildings or processes.

4.7.2 Possible technical problems of operating CHP plants with vegetable oil

The fuel pumps and fuel injection systems are the most critical parts of combined heat and power units operated using vegetable oil. They are the source of most problems. Poor fuel quality or harmful contaminants can lead to a rapid build-up of residue on the injection nozzles and in the fuel system. The repercussions of such debris accumulation are insufficient atomization of the fuel, which can then lead to contamination of the engine oil.

The biggest threat to vegetable oil engines is posed by a failure of the central lubrication system, which usually leads to major engine damage. In addition to harmful contaminants in the fuel, often a gradual thixotropy of the lubrication oil is responsible for this type of damage. For these reasons, oil changes and maintenance checks at least every 250 hours are recommended for vegetable oil engines. Moreover, it is not just an exchange of the engine oil and filter equipment that should be carried out, but also an analysis of the quality of the oil found.

Vegetable oil behaves differently from refined fossil fuels in many aspects. It is therefore also advisable to equip vegetable oil engines with cool-down timers, which guarantee the slow cooling-down of the engine unit. This not only prevents engine damage from overheating but also prevents carbonization of the nozzles and pistons by uncontrolled vegetable oil burning.

Making a prognosis for maintenance intervals for vegetable oil combined heat and power plants is difficult because little operational experience has been gathered. Although previous projects recorded operational successes, heavy setbacks with major engine damage were also suffered. The operation of combined heat and power plants with natural vegetable oil depends on very many parameters, and often on parameters that cannot be sufficiently monitored, with the result that the exact causes of the failures often cannot be determined.

These facts mean there is a need for an increase to maintenance funds of over €0.02 per kWh of electrical energy produced. This should be taken into account at the earliest possible financial planning stage. To ensure the success of such a scheme, as mentioned at the beginning of this chapter, a cooperative partnership with producers and suppliers should always take the form of a demonstration project. Current findings on vegetable oil engines in stationary, long-term operation have shown other approaches to be very risky.

4.8 Project management

4.8.1 General project planning

To implement a CHP project with liquid biofuels, generally speaking the same steps should be taken as required to set up a stationary CHP plant with fossil fuels.

A project plan should proceed according to the following steps:

- Make a preliminary investigation and survey of the current situation.
- Analyse heat requirements at the site.
- Analyse costs and make preliminary investigations into the fuel costs.
- Check economic efficiency.
- Draft CHP plant concepts for various problem scenarios.
- Select a suitable technique.
- Select a suitable operating method.
- Obtain quotes for suitable CHP plants including maintenance costs.
- Pre-plan for the peripheries – design of electrical connections etc.
- Obtain necessary authorizations from the authorities (building laws, emission controls).
- Carry out final examination of economic viability.
- Start the project.

As already described in the introduction to this chapter, so far there has been little practical experience with combined heat and power plants operated using natural vegetable oil. Most successful projects are still running with the technical support of the engine manufacturer or in the framework of scientific studies.

If the reader intends to launch such a project in pure market conditions, sufficient guarantees from the manufacturer, including regarding maintenance costs, should be obtained in advance. Energy projects with natural vegetable oil still have something of a research nature. The success of the project partnership and the scheme should be measured by the contracts closed.

The following section describes the most important elements for the success of an energy project.

4.9 Technical planning

To ensure the success of a project, it is essential to procure and ensure the long-term supply of fuels in sufficient quantities and adequate quality. Long-term and – above all – quality-bound supply contracts with producers of vegetable oil are strongly advised. Commonly they deliver the oil in palette tanks holding up to 1000 litres.

As the vegetable oil delivered has a storage life of about 6 months, the tank logistics should be designed in such a way as to minimize storage times within the plant. In larger CHP plants with high oil consumption and a high annual usage, delivery in a tank lorry and storage in an external fuel tank is possible, although attention must always be paid to making sure that the tanks are free from other residue before use. If other residue does exist it can lead to instant major damage to the engine and the fuel management systems. In practice, adhering to quality guidelines is hard with changing suppliers. The optical and sensory controls proposed by many do not give the complete picture.

In contrast to fossil fuels such as diesel or fuel oil, vegetable oil is a perishable natural product that can be damaged by many external influences. Oxygen, light, heat and heavy metal ions from iron or copper compounds can lead to massive damage to the oil. Here the viscosity of the oil increases and the likelihood of clogging of fuel pipes, injection pumps and nozzles increases exponentially. Such processes can also considerably affect the lubrication capacity of the engine oil.

For storage the following recommendations should be noted:

- Keep the storage temperature below 10°C.
- Only heat the fuel immediately before use.
- Store biofuels in photoresistant tanks.
- Fill up the tank to prevent oxygen getting in.
- Drain condensed water arising from temperature fluctuations.
- Keep out contaminants.
- Do not use metals or metal compounds that can oxidise (such as copper, brass, iron); it is advisable instead to use stainless steel or permanent synthetic materials such as Perbunan or Viton.
- Take a sedimentation zone into account.
- Do not store plant oil for longer than 6 months.

The supply pipes should also be designed according to these criteria, and adequate pipe dimensions are important. Pipes that are intended to transport 30 litres of vegetable oil per hour for a long period must have a minimum diameter of 14 mm. Pipes should be designed in such a way that they can be drained completely, and they should allow access for sampling, and gauge glasses. The pumps used should be suitable for operation with viscous materials; hose pumps are suitable for this. They should also serve short pipes and have double the delivery rate. The installation of a buffer tank in front of the fuel pump, mostly equipped with a return pipe, is advisable. The systems should have high-tech double filters, so that as many dirt particles as possible can be trapped, which prevents any engine damage occurring.

5 Small combustion systems

5.1 Introduction

Chopping wood, flying sparks, ash, dirt, smoke and bad heating performance, these are the things that still come to a lot of peoples' minds when they think about small-scale wood heating. However reality has advanced way beyond this.

Today, small-scale heaters for biomass are technically very advanced. Superior in environmental performance and aesthetics they are able to substitute fully for heating systems based on fossil energy. Through their possibilities for fully automatic feeding, the wide range performance controls and the comfortable computerized operation, most users are unable to detect much difference from standard heating systems. Even the old problem field of smoke emissions has been solved by lambda-controlled secondary firing mechanisms.

In the light of continuous technical advancements in the field of small-scale heating systems, heating with wood is on the increase. Equally comfortable, and environmentally and aesthetically superior, it is starting to pull ahead of fossil heating installations in many ways. Soon wood might be again the first choice of heat energy in modern residential buildings.

5.1.1 Who should read this chapter?

This chapter is written for those who are considering the alternatives of small-scale biomass heating for inclusion in residential building projects. The guide shows many alternatives and with the information provided, a qualified decision on design concepts and specific models can be taken.

5.1.2 What information is provided in this chapter?

The chapter provides all the necessary information for the optimal choice, dimensioning and planning of small-scale wood heating systems for residential applications. Within the chapter all important technical design forms are summarized. This way it is possible to make a quick and application-orientated choice of the right heater type. This chapter provides an extensive overview of the technical possibilities of small-scale wood heating systems and the resulting technical requirements for their application.

5.2 Heat demand of buildings

The most important parameter for selecting and measuring heating systems is the heat demand. This comprises two elements: the space heating demand and the hot water demand. Whereas the hot water demand can be taken to be constant at 12.5 kWh/m^2 living space, the measurement of the space heating demand depends on many parameters. Various strategies for determining this heat demand are described on the following pages.

Table 5.1, based on empirical values, can be used to make a rough approximation of a building's space heating demand.

In many countries there are national or regional technical provisions on heat insulation. These include building regulations that take climate protection into consideration and are designed to reduce the heat demand of new buildings. Based on the respective regulations in accordance with the age of the buildings, it is possible to make a rough estimate of the average heat demand of the existing building stock.

Table 5.1.
Specific heat demand of buildings
Source: www.wamsler-hkt.de

Building	External thermal insulation	Window glass	Room height (m)	Specific heat demand (W/m²)
Old building	No	Single	2.5	190
Old building	No	Single	2.5	160
Old building	Part	Thermopane-glazed	2.5	130
Old building	Part	Thermopane-glazed	2.5	110
New building	Yes	Thermopane-glazed	2.5	90
New building	Yes	Triple-glazed	2.5	70

Technical analyses of the building stock in Germany result in the annual heat demands for different types of buildings shown in Figure 5.1.

Carrying out thermal insulation calculations according to regionally or nationally applicable calculation procedures for individual buildings can often provide a good starting point for establishing the heat demand of a building.

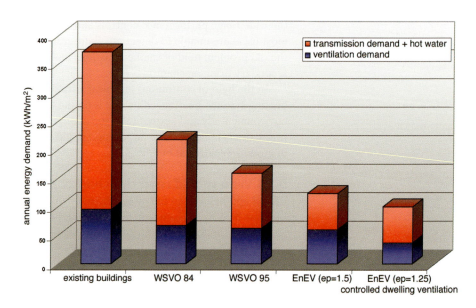

Figure 5.1.
Mean annual heat demand of buildings
Graphic: www.sesolutions.de
Data: Wamsler

Despite the supposed precision of the prescribed calculation procedures, in practice there are often considerable differences between the calculated and the actual heat demand. This is because such procedures are very much simplified and standardised. For instance, most provisions use standardised temperature sequences and user behaviours that, although correct on average, can often depict individual buildings incorrectly owing to a lack of more detailed information.

Figure 5.2 shows the ratio between the required boiler output in kilowatts and the enclosed building space in cubic metres. The range of variance of building stock without thermal insulation measures is marked in red. The range of variance of buildings that are insulated according to the German Thermal Insulation Ordnance (WSchV) from 1995 is marked in yellow. The range of variance of existing low-energy houses or buildings that were constructed according to the current Energy Saving Ordinance (EnEV) is marked in violet.

Figure 5.2 shows that there is a particularly broad divergence of energy requirements with old buildings. The more modern the building, the smaller the variance in the heat requirement permitted according to the provisions.

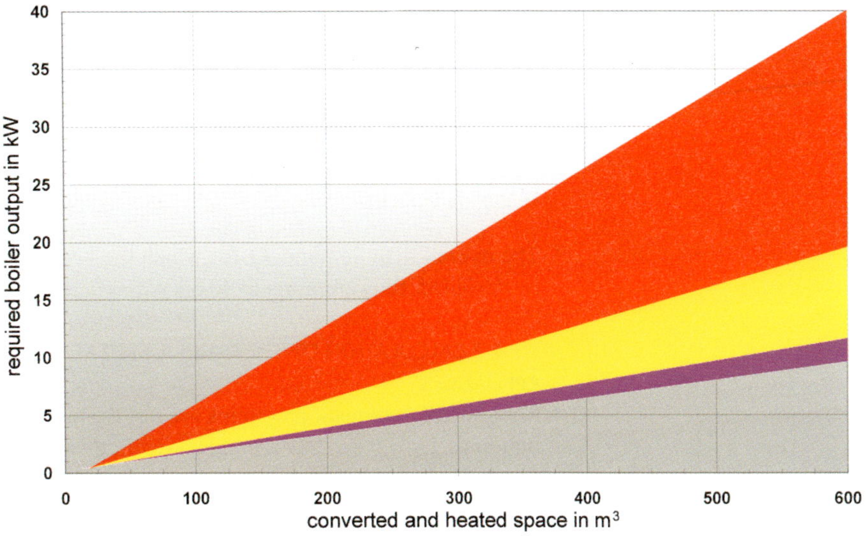

Figure 5.2.
Ratio of boiler output to enclosed space
Graphic: www.sesolutions.de
Data: Wamsler

5.2.1 Detailed measurement of maximum heating output

The maximum heating output required for small combustion systems for central heating systems and heat appliances can be calculated in various ways.

5.2.1.1 CALCULATION OF HEAT DEMAND FOR CENTRAL HEATING SYSTEMS

The following formula allows sufficiently precise measurement of the heat demand, Q, when sizing wood-fired heating systems for buildings:

$$Q = H \times A \times F_1 \times F_2$$

where H is the specific heat demand (Table 5.2), A is the heated living area in m², F_1 is a correction factor for other minimum temperatures (Table 5.3), and F_2 is a correction factor for building type (Table 5.4).

Table 5.2.
Specific heat demand of buildings
Source: www.wamsler-hkt.de

Building	External thermal insulation	Window glass	Room height	Specific heat demand
Old building	No	Single	2.5	190
Old building	No	Single	2.5	160
Old building	Partly	Thermopane-glazed	2.5	130
Old building	Partly	Thermopane-glazed	2.5	110
New building	Yes	Thermopane-glazed	2.5	90
New building	Yes	Triple-glazed	2.5	70

Table 5.3.
Correction factor F_1 for absolute minimum temperature
Source: www.wamsler-hkt.de

T_{min} (°C)	F_1
6	0.76
8	0.82
10	0.88
12	0.94
14	1.00
16	1.06

Table 5.4.
Correction factor F_2 for building type
Source: www.wamsler-hkt.de

Building type	F_2
Detached house	1.00
Terraced house (end house)	0.95
Terraced house (middle)	0.90
Apartment building:	
< 8 dwelling units	0.70
> 8 dwelling units	0.65

Example
For a residential building with six dwelling units, an overall living area of 420 m² and situated in an area with a minimum ambient temperature of 16°C, a specific heat demand H of 130 W/m² living area was determined according to Table 5.2.
How large is the necessary overall heat demand, Q?

$$Q = H \times A \times F_1 \times F_2 = 130 \text{ W/m}^2 \times 420 \text{ m}^2 \times 1.06 \times 0.70 = 40.5 \text{ kW}$$

The overall heat demand, Q, for which the heat output of the heat generator should be designed, thus amounts to 40.5 kW.

5.2.1.2 CALCULATION OF THE MAXIMUM HEAT OUTPUT REQUIRED FOR HEARTH APPLIANCES

Many small combustion systems are not designed to supply the entire heat for buildings. However, they are able to fulfil diverse heating tasks for individual rooms or areas within buildings.

If it is intended to size such hearth appliances, detailed knowledge is required of the heat demands for the respective spaces. Here it is not possible to implement global, building-related approaches, as in most cases there is considerable variation in the heat demands. Representative of other effects, Figure 5.3 illustrates the differing heat demands depending on the position of a room within a building.

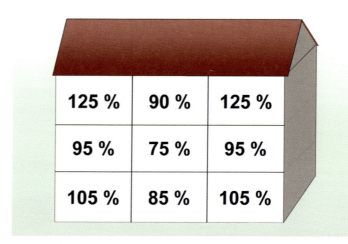

*Figure 5.3.
Variation of mean heat demand in buildings
Graphic: www.sesolutions.de
Data: Wamsler*

If you know the maximum heat output required for a building and the position of the room to be heated with a hearth appliance, it is possible to approximate the space heating demand with the percentage values given in Figure 5.3.

In many cases this method is sufficiently precise for selecting the appropriate heat output of hearth appliances.

Note: If there is doubt as to whether there is sufficient heat output, it is better to choose a larger model. At the same time, however, it should be ensured that hearth appliances are not installed that are unnecessarily large; otherwise the efficiency throughout the year is reduced, resulting in higher fuel consumption and emissions.

5.2.1.3 ALTERNATIVE PROCEDURES FOR CALCULATING MAXIMUM HEAT OUTPUT REQUIRED FOR HEARTH APPLIANCES

Should there be no reliable calculation of the necessary heat output, it can be determined using a points system. Table 5.5 must be worked through and the points added together if a statement is applicable.

If the points total for a space or a building type is known (not insulated, partly insulated, or low energy), it is possible to determine the necessary heat output of the hearth appliance using the graphs in Figures 5.4–5.6.

Table 5.5.
Points table for calculating maximum
heat output output
Source: www.wamsler-hkt.de

Evaluation factor	Points
Detached house	1
Attic room	2
Room with two unheated internal walls	1
Room with three unheated internal walls	2
Uninsulated room with external-facing walls or ceilings	2
Room adjoining or above driveway	1
Each wall of the room is an external wall	2
Windows larger than 1/5 of the room's external area	2
Room orientation NW – N – NE – E	1
More than 600 m above sea level or particular cold area	1
Particularly exposed to wind	2
Room temperature must be more than +20°C even if very cold outside	1
Frequently used room (store room or bar room)	2

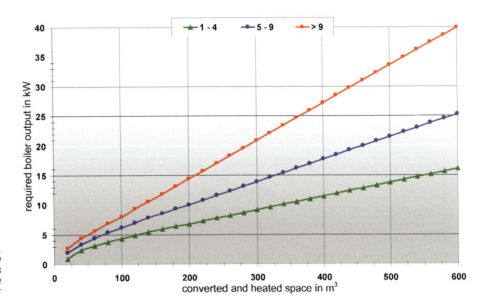

Figure 5.4.
Ratio of boiler output to enclosed space
for uninsulated spaces
Graphic: www.sesolutions.de
Data: Wamsler

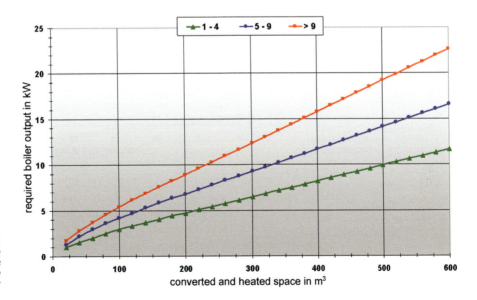

Figure 5.5.
Ratio of boiler output to enclosed space
for conventionally insulated spaces
Graphic: www.sesolutions.de
Data: Wamsler

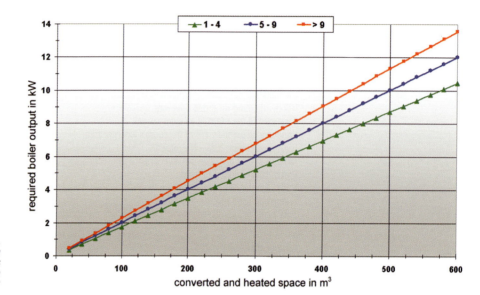

*Figure 5.6.
Ratio of boiler output to enclosed space
for low-energy homes
Graphic: www.sesolutions.de
Data: Wamsler*

If the maximum necessary output of a hearth appliance is known, the wood demand for the respective output can be determined from Figures 5.7 and 5.8. The difference in the energy content of natural wood and artificially compressed wood pellets is a result of their different storage densities. Figure 5.7 shows the change in the daily wood demand with ambient temperature. Figure 5.8 enables an approximation to be made of the overall annual fuel demand.

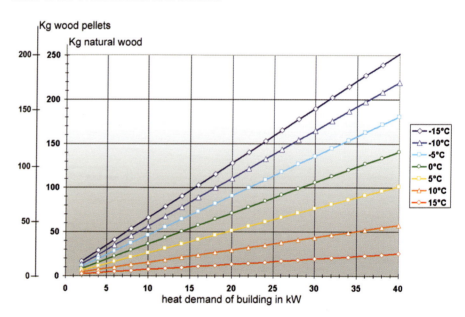

*Figure 5.7.
Daily wood demand/heat output
Graphic: www.sesolutions.de
Data: Wamsler*

The annual and daily wood demands are important criteria when choosing hearth appliances. If you do not choose an automatic boiler, such as a central pellet or woodchip boiler, you will always have to feed the wood manually.

For wood logs this means you will generally have to prepare, split and dry the wood yourself. When purchasing wood logs, briquettes and wood pellets in sacks, the fuel still has to be fed into the heat generator and 1–3% of its weight will have to be disposed of as ash.

Should a manually fed wood heating source be chosen as the only heat source for a house, the aspects of procuring and stoking should be discussed prior to installation. When consulting the customer it should be pointed out that there will be exceptional situations such as during illnesses or holidays.

Note: The installer offering advice should openly discuss the advantages and disadvantages of manual feeding to ensure that the customer will remain satisfied

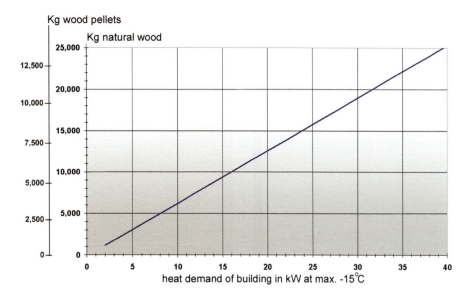

Figure 5.8.
Annual heat demand/heat output
Graphic: www.sesolutions.de
Data: Wamsler

later. The consultant should not worry that she might lose the customer to another energy source such as oil or gas, as wood heating provides enough technical possibilities for providing fully automatic heat.

In addition, heating with wood is still very much an experience to be enjoyed. This environmentally friendly form of heating also offers numerous other less tangible advantages, such as a cosy ambience and a traditional feel (Figure 5.9).

Figure 5.9.
Heating with wood: tradition and lifestyle
Photo: www.hase.de

5.2.2 Seasonal distribution of annual heat demand

Hot water demands for domestic appliances are generally sporadic events. You can best cover these peak heat demands by using sufficiently sized storage tanks (accumulators). The peak heating capacity of a heat generator, however, is generally needed only during the few days of the year when there are continually very low temperatures. If the peak output of a boiler is sized too small, the room temperature of the building cannot be maintained at the desired level.

The actual course of a building's heat demand without domestic hot water heating being taken into consideration can be depicted in an annual load duration curve (Figure 5.10). Here the percentage proportion of the recorded heat demand is plotted relative to the 8760 hours of the year.

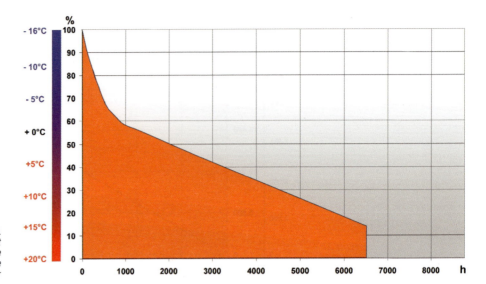

Figure 5.10.
Annual heat demand relative to ambient temperature
Graphic: www.sesolutions.de
Data: Wamsler

The shape of the curve is derived from the characteristic temperature course for each year. The steep curve on the left of the graph depicts the few extreme winter days in a year in central Europe. The long, flat slope on the right of the graph depicts the gradual temperature fluctuations caused by the changing seasons.

The heating period in central Europe is defined as ending after 6500 hours of the year; after this point there is no further heating demand.

In order to calculate and specify design variants for single and combined heat generators, it is essential to know the monthly distribution of the annual heat demand. The mean monthly and accumulated distributions of the annual heat demand are provided in Tables 5.6 and 5.7.

Heating systems are not active outside the heating period. This means that there is no work in terms of fuel management. When procuring fuel, and sizing the fuel storage for heat generators, it is necessary to know the distribution of the heat demand across the heating period. This is shown in Table 5.7.

These basic data enable designers and building owners to make the optimum choice for a building, and possibly combine different heat generators. Generally, two variants are available: monovalent and bivalent systems.

Table 5.6.
Distribution of the annual heat demand across the entire year
Source: www.wamsler-hkt.de

Month	Monthly proportion (%)	Cumulative proportion (%)
January	17.7	17.7
February	15.1	32.8
March	13.4	46.2
April	9.5	55.7
May	2.9	58.6
June		58.6
July		58.6
August		58.6
September	1.9	60.5
October	9.5	70.0
November	13.2	83.2
December	16.8	100.0

Table 5.7.
Distribution of the annual heat demand across the heating period
Source: www.wamsler-hkt.de

Month	Monthly proportion (%)	Accumulated proportion (%)
September	1.9	1.9
October	9.5	11.4
November	13.2	24.6
December	16.8	41.4
January	17.7	59.1
February	15.1	74.2
March	13.4	87.6
April	9.5	97.1
May	2.9	100.0

- With monovalent systems, the heat generator is required to cover the entire heat demand of the building, even in the harshest winters. For this reason, such heat generators must be designed to meet the calculated maximum heat demands of the building.
- With bivalent systems, the heat demands of a building are covered by two or more different types of heat generator. A common example of a bivalent system is the coupling of pellet boilers with solar plants.

Whereas monovalent systems by their very nature work alone, bivalent heating systems are differentiated into two types: bivalent alternative systems and bivalent parallel systems

Bivalent alternative systems are operated on an 'either/or' basis. This means that one heat generator must be designed as a monovalent system that can supply the full heat load. The second system is generally designed with a smaller output. The systems are operated alternately, so that the system with the lower output supplies the building in warmer periods and the larger system supplies the heat in colder periods. An example of this is the combination of pellet stoves with central heating boilers.

Bivalent parallel systems are operated on an 'and' basis, whereby both systems cover the heat requirement according to their individual capacities. If bivalent parallel systems are to work efficiently, it is essential that there is an electronically controlled storage tank.

5.2.2.1 OPERATING STRATEGIES FOR BIVALENT ALTERNATIVE SYSTEMS

Figure 5.11 shows the distribution of the heat supply to two system parts (orange, green) using bivalent alternative systems.

The main heat generator with the peak output is operated for only 1400 hours a year during the coldest part of the heating period, when temperatures are below 0°C. Above this temperature the second heat generator takes on the full heating supply of the building. Examples of such bivalent alternative systems include the transfer from a central heating boiler to a pellet-fired or wood-fired stove.

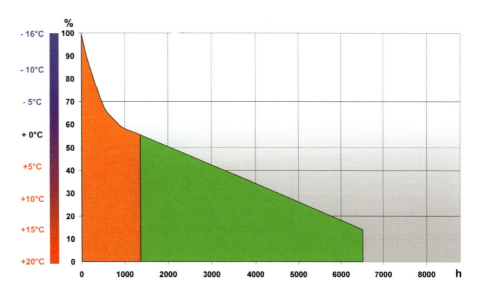

Figure 5.11.
Operating strategy for bivalent alternative systems
Graphic: www.sesolutions.de
Data: Wamsler

5.2.2.2 OPERATING STRATEGIES FOR BIVALENT PARALLEL SYSTEMS

Figure 5.12 shows the distribution of the heat supply to two system parts (orange, green) using bivalent parallel systems.

The ancillary heat generator is required to supply the peak output only. It only operates for 1400 hours a year during the coldest part of the heating period, when temperatures are below 0°C. Above this temperature the main heat generator with the lower basic load provides the full heating supply for the building.

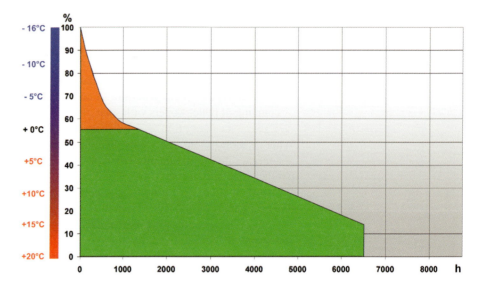

Figure 5.12.
Operating strategy for bivalent parallel systems
Graphic: www.sesolutions.de
Data: Wamsler

Examples of such bivalent parallel systems include the combination of district heating systems with peak load boilers, and the integration of solar energy with wood-fired heating systems.

5.3 Choosing small combustion systems for heating buildings

There are numerous alternatives for providing buildings with CO_2-neutral, wood-fired heating. Here it is possible to deploy not only central heating systems but distributed hearth appliances in combination with solar energy or other energy sources. The combination possibilities are numerous, and offer a suitable output for every application area.

Ultimately, the choice of small combustion system is determined by technical parameters, whereby it is necessary to have precise knowledge of all the qualities of the heat generators. The heat generators described in this section are:

- open fireplaces
- closed fireplaces
- wood stoves
- pellet stoves
- central heating cookers
- tiled stoves
- log-fired central heating boilers
- central pellet boilers
- woodchip boilers
- combination boilers.

The initial selection of the heat generator is always based on the required maximum output determined by the heat demand calculation. If it is intended to use a heat generator in a bivalent, parallel operating system, it is possible to use almost any combination of distributed and centralized heat generators.

When using a combination of various heat generators, it is possible to choose a small combustion system with an output that is too small. However, a solid fuel-powered heat generator should never be operated beneath its intended performance class; otherwise the efficiency drops, and this can lead to considerable emissions. Figure 5.13 shows small combustion systems classified according to their heat output.

In addition to the heat output, the annual heat demand is also an important selection criterion. Not every installation that has an appropriate output is suitable for covering every annual heating demand. Figure 5.14 shows the same small combustion systems classified according to the annual heat demand that they can supply.

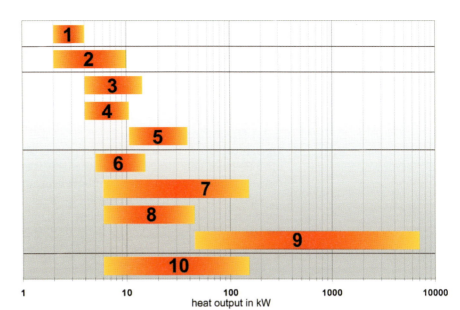

*Figure 5.13.
Peak output ranges for small combustion systems
Graphic: www.sesolutions.de*

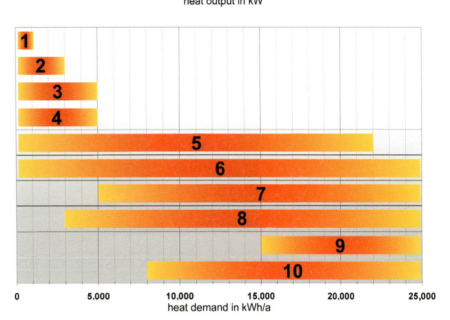

*Figure 5.14.
Annual output that can be provided by small combustion systems
Graphic: www.sesolutions.de*

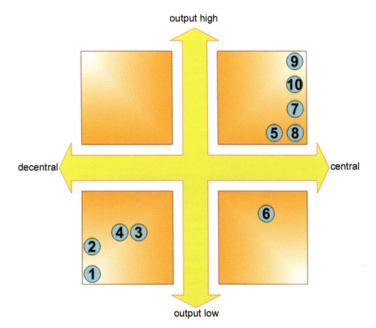

*Figure 5.15.
Selection criteria for output and decentralization of small combustion systems
Graphic: www.sesolutions.de*

A third selection criterion is the possibility of connecting to a central heating system. Although not all buildings have such a centralized system, not all small combustion systems can be connected to such a system when they do exist, as is shown in Figure 5.15.

Ultimately, however, personal taste and individual comfort play a central role when selecting the system types illustrated in the following sections.

5.3.1 Open fireplaces

Open fireplaces do not actually have a positive heating effect. If wood is burned in an open fireplace, only around 20% of its energy is used for space heating in the form of radiation. The remainder escapes without being used through the chimney.

Open fireplaces are used mainly for increasing the living quality, not for heating purposes. For this reason they should be used only occasionally, for instance during the transitional period leading up to the heating period.

Figure 5.16.
Open fireplace
Photo: www.kamikassette.de

5.3.1.1 TECHNICAL SPECIFICATIONS FOR OPEN FIREPLACES
See Table 5.8.

Table 5.8.
Technical specifications for open fireplaces
Data: www.sesolutions.de

Primary application	Visual effect in room, cosiness
Installation location	Living space
Heating area	Living space (limited)
Heat emission through radiation	Yes
Heat emission through convection	No
Heat emission through water heat exchanger	No
Type of combustion	Room-air-dependent
Combustion chamber	Open
Output range (heat)	1–3 kW
Efficiency	< 20%
Usable fuel	Wood logs, wood briquettes
Ignition process	Manual
Flue temperature	Approximately 180°C
Joint use of chimney with other systems	No
Required chimney diameter	To be calculated individually
Chimney soot fire resistance	1200°C
Moisture-resistant chimney	Yes
Prefabricated heating system	No
Fresh air demand	0.036 m^3/h per cm^2 opening
Safety clearance to combustible material at the front	> 1.00 m
Safety clearance to combustible material at the sides	> 0.30 m
Safety clearance to combustible material above	> 0.70 m
Safety clearance to combustible material on the floor	Fireproof base

5.3.1.2 STRUCTURAL REQUIREMENTS FOR OPEN FIREPLACES

An open fireplace requires its own separate chimney and fresh air intake. No other hearth appliances should be operated in rooms with open fireplaces. These rooms should also not be equipped with extraction hoods. This prevents any possible emission of smoke back into the living spaces, which could lead to people being poisoned.

In contrast to all other hearth appliances, open fireplaces have an open combustion chamber facing the living space. The rear and side walls of open fireplaces are either manufactured from precast fireclay masonry or are constructed using prefabricated components.

If it is intended that the fireplace should be used during the transitional period for space heating, the necessary heat opening can be calculated:

Room volume in m^3 × 30 = necessary fireplace opening in cm^2

Example
A 100 m^3 room (40 m^2 floor space with a room height of 2.50 m) would require a fireplace opening of 3000 cm^2. This corresponds to a fireplace opening of 50 cm × 60 cm.

At the same time, the fireplace opening must have a balanced ratio to the volume of the living space. The maximum permitted fireplace opening can be calculated using the following formula:

Room volume in m^3 × 60 = maximum opening of the fireplace in cm^2

Figure 5.17 shows recommended dimensions for fireplace openings in relation to room size. However, overriding these are the legally binding provisions for designing open fireplaces and the requirements of the relevant local authorities.

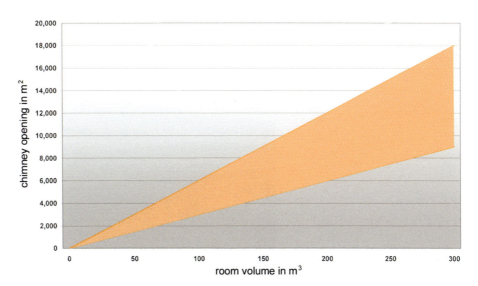

Figure 5.17.
Ratio chimney opening/room size
Graphic: www.sesolutions.de

As well as the design of the fireplace's combustion opening, the supply of fresh air to the fireplace is also important. A fireplace requires a minimum air change of 0.036 m^3/h per cm^2 of combustion chamber opening:

Fireplace opening in cm^2 × 0.036 = permanent fresh air intake in m^3/h

Example
The operation of a fireplace with an opening of 3000 cm^2 requires a permanent fresh air intake of more than 108 m^3/h.

However, windows and doors in modern houses are so well sealed that it is impossible for such air volumes to flow in naturally. Hence the combustion air must be fed in via a sufficiently sized fresh air duct from outside.

5.3.1.3 TECHNICAL PROBLEMS AND SOLUTIONS FOR OPEN FIREPLACES
See Table 5.9.

Table 5.9.
Most frequent causes of problems with open fireplaces
Data: www.sesolutions.de

Problems firing up	
Wood does not ignite	
Causes:	
Wood is too thick	Chop the wood logs into smaller pieces
Wood is too damp	Dry out the wood for at least 1 year
Too little air supply	Use bellows, control room air supply
Operational problems	
Wood burns without bright yellow flame, smoulders or even goes out	
Causes:	
Wood is too damp	Dry out the wood for at least 1 year
Too little air supply	Check flue draught, have chimney cleaned
Too much soot forms, the chimney breast becomes black	
Causes:	
Wood is too damp	Dry out the wood for at least 1 year
Too little air supply	Check flue draught, control room air supply
Too little wood	Put more wood on the stove to create a firebed
The chimney becomes moist and sooty, condensation drips down from the chimney	
Causes:	
Wood is too damp	Dry out the wood for at least 1 year
Flue gas is too cold	Heat up chimney with maximum output
Chimney cross-section is incorrect	Refurbish chimney

5.3.2 Closed fireplaces

If a fireplace is enclosed behind a glass door or glass pane, then it is described as a closed fireplace (Figure 5.18). These are also often known in the trade as built-in fireplaces or, if they are factory built, as zero-clearance fireplaces. A special kind of zero-clearance fireplace is an insert. This comprises a firebox with ash pans, flue gas collector, hot gas tubes and flue pipes.

Figure 5.18.
Closed fireplace
Photo: www.kamikassette.de

In contrast to open fireplaces, the firebox is closed. This allows improved control of the combustion air inlet. This increases the firebox temperature and leads to a perceptible increase in efficiency and combustion quality.

However, because the geometry of the chimney and the combustion remain essentially identical, this does not increase the air change in the room. Therefore some closed fireplaces are additionally equipped with convection ducts and warm air tubes.

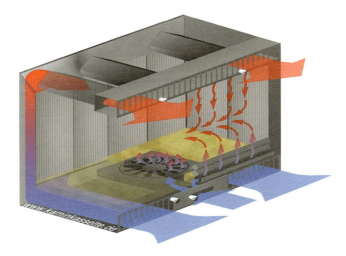

Figure 5.19.
Air circulation in a closed fireplace
Graphic: www.kamikassette.de

Depending on the size and the manufacturer, inserts have a thermal output of between 5 kW and 10 kW. The heat is emitted by radiation into the room. Inserts are suitable for heating rooms during transitional periods.

Those who, despite the poor overall efficiency, wish to heat their rooms with inserts or closed fireplaces, should consult the particular manufacturer of the products about its implementation and individual sizing. This will also specify the minimum air change that will need to be provided in individual cases.

5.3.2.1 TECHNICAL SPECIFICATIONS FOR CLOSED FIREPLACES
See Table 5.10.

Table 5.10.
Technical specifications for closed fireplaces
Data: www.sesolutions.de

Primary application	Cosiness, space heating
Installation location	Living space
Heating area	Living space (limited)
Heat emission through radiation	Yes
Heat emission through convection	Yes
Heat emission through water heat exchanger	No
Type of combustion	Room-air-dependent
Combustion chamber	Closed
Output range (heat)	5–10 kW
Efficiency	< 40%
Usable fuel	Wood logs, wood briquettes
Ignition process	Manual
Flue temperature	Approximately 400°C
Joint use of chimney with other systems	No
Required chimney diameter	To be calculated individually
Chimney soot fire resistance	1200°C
Moisture-resistant chimney	Yes
Prefabricated heating system	Inserts yes, other closed fireplaces no
Fresh air demand	To be calculated individually
Safety clearance to combustible material at the front	> 0.80 m
Safety clearance to combustible material at the sides	> 0.30 m
Safety clearance to combustible material above	> 0.70 m
Safety clearance to combustible material on the floor	Fireproof base

5.3.2.2 STRUCTURAL REQUIREMENTS FOR CLOSED FIREPLACES
Almost any open fireplace can be converted with an insert to form a closed fireplace. This installation should generally be reported, however, to the responsible local authorities or to a certified chimney sweep. They will check that the fireplace and chimney are able to withstand the increased temperatures resulting from the closed method of construction.

Each fireplace normally has its own individual geometry. For this reason, it is always essential to check the installation on an individual basis; it is impossible to make any general statements as to suitability. The following criteria should be taken into account when considering the installation of an insert:

- There should be no combustible material in the radiation area.
- The apron flashing for the chimney must be of fireproof, inflammable material.
- There should be sufficient insulation around the fireplace surrounds.

Only a qualified specialist firm can clarify these points beyond doubt, and carry out the necessary technical consultations with the responsible local authorities.

Professionally installed inserts are capable of providing cosy, clean and spark-free space heating in existing open fireplaces. Inserts are fed manually with firewood. Closed fireplaces can be heated with wood logs or wood briquettes. As with the stoking, the ash must be removed manually from the ash pan. For these reasons, and because of their low overall efficiency, inserts are only suitable for heating living spaces during transitional periods.

5.3.2.3 TIPS FOR REMEDYING OPERATING PROBLEMS WITH CLOSED FIREPLACES
See Table 5.11.

Table 5.11.
Most frequent causes of problems with closed fireplaces
Data: www.sesolutions.de

Operational problems	
Wood burns without bright yellow flame, smoulders and even goes out	
Causes:	
Flue draught is too weak	Install draught regulator, clean chimney connection
Wood is too damp	Dry out the wood for at least 1 year
Too little air supply	Open air regulating damper, control room air supply
Too much soot forms, the viewing window does not remain clean	
Causes:	
Wood is too damp	Dry out the wood for at least 1 year
Too little air supply	Open air regulating damper, control room air supply
Too little wood	Put more wood in the combustion chamber
Although the fire burns strongly, the insert does not become warm	
Causes:	
Flue draught is too strong	Install draught regulator
Smoke escapes through the front door	
Causes:	
Front door is not closed	Make sure front door is closed tightly
Too little air supply	Open air regulating damper, control room air supply
Chimney cross-section is too narrow	Widen chimney
Too much soot in chimney	Contact chimney sweep
Wind-induced downdraught from chimney	Install chimney cap

5.3.3 Wood stoves

The wood stoves available these days are technically improved versions of the well-known room stoves from the past. They are freestanding, and connected to the chimney via a sealed flue pipe. Wood stoves have airtight, lockable front doors that normally have quartz glass panes to enable you to view the fire (Figures 5.20 and 5.21).

Wood stoves emit most of their heat by radiation from the heated surface. Many types also have a convection jacket that allows cold air to be drawn in around the stove before being released again as warm air via air slots in the upper part of the stove.

Wood stoves generally weigh between 13 kg and 26 kg per kW of thermal output, so that an overall weight of 40–80 kg can be expected per square metre of stove heating area. Modern stoves have a fan that regulates the heat output. Depending on the model and manufacturer, it is possible to regulate the output of these stoves between 2 kW and 15 kW. Some manufacturers also provide stoves with remote controls.

Figure 5.20.
Freestanding wood stoves
Photo: www.hase.de

Figure 5.21.
Modern wood stove
Photo: www.hase.de

5.3.3.1 TECHNICAL SPECIFICATIONS FOR WOOD STOVES
See Table 5.12.

Table 5.12.
Technical specifications
for wood stoves

Primary application	Space heating, cosiness
Installation location	Living space
Heating area	Living space
Heat emission through radiation	Yes
Heat emission through convection	Yes
Heat emission through water heat exchanger	Depends on construction type
Type of combustion	Room-air-dependent
Combustion chamber	Closed
Output range (heat)	3–15 kW
Efficiency	< 90%
Usable fuel	Wood logs, wood briquettes
Ignition process	Manual, automatic
Flue temperature	180–200°C
Joint use of chimney with other systems	Yes
Required chimney diameter	To be calculated individually
Chimney soot fire resistance	1200°C
Moisture-resistant chimney	Yes
Prefabricated heating system	Yes
Fresh air demand	4 m^3 space volume/ kW
Safety clearance to combustible material at the front	> 0.80 m
Safety clearance to combustible material at the sides	> 0.20 m
Safety clearance to combustible material above	> 0.70 m
Safety clearance to combustible material on the floor	Fireproof base

5.3.3.2 STRUCTURAL REQUIREMENTS FOR WOOD STOVES

5.3.3.2.1 Installation

Stoves build up heat when operated. This is emitted as radiation through the viewing window, or increases the temperature of the external surface. For this reason such stoves must always be kept at a safe distance from combustible building materials, furniture, curtains or other decorative fabrics.

When designing the installation of stoves in living and workrooms, specific minimum clearances and provisions must be observed. For instance, there must be a minimum clearance of 80 cm around the viewing window. Technical measures such as ventilated radiation protection mean that this distance can sometimes be reduced in individual cases. Nevertheless, the individual requirements of the stove manufacturer must always be observed.

There is less risk of surrounding materials catching fire with the other areas of the stove, such as the rear wall, side cladding and stovepipe. Therefore the distance to inflammable objects can be reduced to 20 cm from the stove.

The stove should not be placed directly on combustible floors such as carpets, parquet or cork. It always requires a base made of a non-combustible building material such as ceramic tiles, stone, glass or steel. As a rule this base should extend 50 cm beyond the front of the stove and 30 cm on either side to ensure optimum fire protection. The individual requirements of the manufacturer are also decisive here when designing the fireproof base.

The most important elements of a stove are the air inlet and circulation openings. The air must be able to pass unobstructed through these at all times. These openings must be kept open, otherwise poor combustion could lead to carbon monoxide poisoning or even the risk of explosion.

If these safety clearances are taken into consideration during planning, this will ensure that the customer is not disappointed later on should it not be possible to fulfil individual architectural wishes through having to meet important safety requirements.

5.3.3.2.2 Fresh air supply

Stoves are classed as room-air-dependent hearth appliances. This also applies if the stove is equipped with a separate external air inlet. There must be a balanced ratio between the room size and the output of the stove.

Generally a space volume of at least 4 m^3 is required per kW of heat output from the stove. For safety reasons, this must not be reduced; if this value is exceeded by too much, the appliance will provide insufficient output in the far corners of the room.

The air supply to a stove is often generated using a noiseless variable-speed fan. The air is drawn in via a central inlet pipe; if an external air inlet is used, this enables wood stoves to be operated largely independently of the room air.

This means of operation is particularly important when using controlled ventilation for living spaces. Smaller amounts of air are then extracted from the space around the stove only to keep ash off the stove's window.

Figure 5.22 illustrates the air circulation within a wood stove. The room air (blue) entering via the fresh air duct is separated into primary and secondary air components (both red). The primary air rises through the ash grate at the base of the stove and oxidises the loaded fuel while giving off heat. The separated secondary air is required for optimum burning of the generated wood gases. It is fed in from above and, after reacting, leaves as flue gas (green) through the stovepipe to the chimney.

With some stoves the hot flue gas also flows through a heat exchanger that enables the heat generated by the stove to be transferred to other rooms using circulating water. The air current that is marked on the window and is directed vertically downwards (red) provides an air washing system to clean the window automatically.

The safety limit for operating wood stoves in a room is a maximum existing negative pressure of 4 Pa. For this reason it is normally forbidden to operate devices that are capable of creating higher negative pressures in the room in which the stove is installed. In particular, these devices include extraction hoods and ventilation systems. Technical measures should be used to ensure that no parallel operation is possible, such as mutual locking of the circuits for the stove and extraction hood.

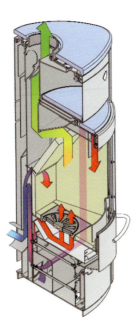

Figure 5.22.
Air circulation in a wood stove
Graphic: www.hase.de

Alternatively, a sufficiently large air inlet opening should be provided in the affected room, whose effectiveness must be proven through measurements.

5.3.3.2.3 FLUE GASES

The maximum output of the boiler is limited by the capacity of the chimney. For this reason, the amount of combustion air drawn in must not lead to positive pressure in the combustion chamber. This means that the suction effect of the chimney must extend as far as the combustion chamber. If this is not the case, either the output of the boiler must be reduced or the chimney must be refurbished.

For extracting flue gases from wood stoves it is necessary to have a chimney that is approved for heating solid fuels. This must have a soot fire resistance of 1200°C. Normally the flue gases have a temperature between 150°C and 200°C, but the flue temperatures in the chimney can drop below 160°C. This can cause condensation, which can damage the chimney.

The stove flue pipe should be designed so that it is not too long. It should always extend vertically from the stove and be less than 2 m in length. For design reasons, and to achieve optimum flue gas extraction, it is advisable to use the original accessories supplied by the stove manufacturer.

Ideally, the chimney should be constructed of damp-proof material and have good heat insulation. This applies particularly for the passage of the chimney through the attic of a house. For operating stoves, most boiler manufacturers specify a minimum chimney diameter of around 13 cm.

Figure 5.23.
Wood stove with bent flue pipe
Photo: www.hase.de

5.3.3.3 TIPS FOR REMEDYING THE MOST FREQUENT OPERATING FAILURES OF WOOD STOVES
See Table 5.13.

Table 5.13. Most frequent operating problems with wood stoves. Data: www.sesolutions.de

Problems firing up	
Wood does not ignite	
Causes:	
Wood is too thick	Chop the wood logs into smaller pieces
Wood is too damp	Dry out the wood for at least 1 year
Too little air supply	Open air regulating damper, control room air supply
Operational problems	
Wood burns without bright yellow flame, smoulders and even goes out	
Causes:	
Flue draught is too weak	Install draught regulator, clean chimney connection
Wood is too damp	Dry out the wood for at least 1 year
Too little air supply	Open air regulating damper, control room air supply
Too much soot forms, the stove walls do not stay clean	
Causes:	
Wood is too damp	Dry out the wood for at least 1year
Too little air supply	Open air regulating damper, control room air supply
Too little wood	Put more wood in the stove
Although the fire burns strongly, the insert does not become warm	
Causes:	
Flue draught is too strong	Install draught regulator
Wood burns too quickly with low heat output	
Causes:	
Flue draught is too strong	Install draught regulator
Wood is chopped into pieces that are too small	Insert larger pieces
Stove operation is incorrect	Read instructions for use
Smoke escapes through the stove door	
Causes:	
Front door is not closed	Make sure front door is closed tightly
Too little air supply	Open air regulating damper, control room air supply
Damper is closed	Open damper
Chimney cross-section is too narrow	Widen chimney
Too much soot in chimney	Contact chimney sweep
Wind-induced downdraught from chimney	Install chimney cap
The chimney becomes moist and sooty, condensation drips down from the chimney	
Causes:	
Wood is too damp	Dry out the wood for at least 1 year
Flue gas is too cold	Heat up chimney with maximum output
Chimney cross-section is too large	Refurbish chimney

As can be seen, most problems that occur when operating wood stoves result from the use of poor-quality fuel, or from insufficient maintenance of the chimney. They can be solved by using dry, optimally chopped fuel and by maintaining the stove correctly.

5.3.4 Pellet stoves

Stand-alone pellet stoves are essentially identical to wood stoves in term of their installation, performance class and connection to the chimney (Figure 5.24). They are also used for the visible provision of heat in living spaces with the possibility of connecting the stove to the central heating system.

In technical terms, however, pellet stoves are substantially different from wood stoves because, apart from the fan, they also provide automatic metering and feeding of the pellets from a storage hopper (Figure 5.25). This enables the stoves to be fed with up to 2 days of heating supplies.

Pellet stoves are used mainly for heating individual living spaces. They are similar to wood stoves installed in domestic living spaces, in terms of both their use and their heating behaviour. Similar to the thermal output of wood stoves, pellet stoves provide heat by radiating heat through the window and by allowing warm air to convect through vents or grills in the stove.

Figure 5.24.
Modern pellet stove
Photo: www.rika.at

Figure 5.25.
Hopper for a pellet stove
Photo: www.rika.at

Pellet stoves are manufactured with a range of heat outputs up to around 11 kW. Built-in fans and standardised fuel enable the output of most pellet stoves to be varied very easily. For instance, pellet stoves can be electronically throttled back to around 30% of their maximum output without showing any notable increase in exhaust emissions.

A further unique technical feature of pellet stoves is the automatic ignition of the fuel. Some manufacturers even enable stoves to be switched on by remote control using a mobile telephone (Figure 5.26).

The most practical application of individual pellet stoves is in restoring and renovating old buildings – that is, in applications for which it would be too costly to provide retrofitted central heating. In most cases there is still a functioning chimney in the building, and this enables pellet stoves to be installed without any problems.

A further application of pellet stoves is in passive and low-energy homes in which the heat demand is so low that such stoves are able to provide the complete heat supply. In these buildings pellet stoves can be operated as warm air/radiation stoves and, combined with a hot water exchanger, as central heating boilers.

Figure 5.26.
Innovative stove control via mobile phone
Graphic: www.rika.at

5.3.4.1 TECHNICAL SPECIFICATIONS FOR PELLET STOVES
See Table 5.14.

Table 5.14. Technical specifications for pellet stoves
Data: www.sesolutions.de

Primary application	Space heating, cosiness
Installation location	Living space
Heating area	Living space
Heat emission through radiation	Yes
Heat emission through convection	Yes
Heat emission through water heat exchanger	Depends on construction type
Type of combustion	Room-air-dependent
Combustion chamber	Closed
Output range (heat)	3–11 kW
Efficiency	< 90%
Usable fuel	Wood pellets
Ignition process	Automatic
Flue temperature	150–200°C
Joint use of chimney with other systems	Yes
Required chimney diameter	To be calculated individually
Chimney soot fire resistance	1200°C
Moisture-resistant chimney	Yes
Prefabricated heating system	Yes
Fresh air demand	4 m^3 space volume/ kW
Safety clearance to combustible material at the front	> 0.80 m
Safety clearance to combustible material at the sides	> 0.20 m
Safety clearance to combustible material above	> 0.70 m
Safety clearance to combustible material on the floor	Fireproof base

5.3.4.2 STRUCTURAL REQUIREMENTS FOR PELLET STOVES

The structural requirements for pellet stoves are similar to those of wood stoves. This applies both to the discharge of the flue gases and to the safe distances of the stoves from flammable furnishings.

Thanks to their integrated storage hoppers, pellet stoves do not require any external storage areas for the fuel. This does not just make more living space available, it also very convenient. As the pourable wood pellets are available not just in 25 kg sacks but in handy small packages of 2.5 kg, it is easy to refill the stoves. They can also be used by older people without any problems.

5.3.4.3 MAINTAINING PELLET STOVES

In contrast to wood stoves, from which the ash must be removed manually after every heating, pellet stoves are designed for semi-automatic operation. Nevertheless, a correctly operated pellet stove still requires regular maintenance and servicing.

The burner of the pellet stove should be checked daily, when operated, for slag and clinkers. Should there be any, they must be removed using the ash rake. The pellet stove's air openings should also be cleaned if necessary. It is important here to prevent any ash from being pushed back into the burner, as this could lead to blockages.

After burning 50 kg of fuel, a pellet stove should undergo the following maintenance programme:

- Switch off the stove and let it cool.
- Remove the burner pot or tray; check for slag or clinkers, and clean all combustion openings.
- Vacuum or brush out the burner pot or tray holder.
- Empty and brush out the ash pan or ash pot.
- Clean the window. Here a good tip is to clean it first with damp newspaper and then polish it with dry newspaper.
- Insert the burner pot or tray and ensure that it is positioned correctly.

For the external care of the stove it is important that the heatproof stove paint used does not generally provide corrosion protection. The stove's metal components can become rusty if they are exposed to severe moisture, for example through wet floors or exposure to damp air.

If rust impairs the external appearance of the stove, the surface should be treated with abrasive paper and then painted over with the original stove paint from the manufacturer. The use of other conventional paints and colours is not recommended, owing to the high temperatures attained.

5.3.4.4 SERVICING PELLET STOVES

Because pellet stoves contain a considerable number of mechanical parts, they should be serviced once a year or after using around 15,000 kg of fuel. Most manufacturers offer comprehensive service contracts. This factory servicing includes complete cleaning of the window and the stove's entire technical equipment. Often, as part of this servicing, switches and ignition elements are replaced or checked to see that they function correctly. For this reason, the servicing provides an important part of the accident prevention and safety precautions, and should on no account be neglected.

5.3.4.5 TIPS FOR REMEDYING THE MOST FREQUENT OPERATING FAILURES OF PELLET STOVES

See Table 5.15.

Table 5.15.
Most frequent operating problems with pellet stoves
Data: www.sesolutions.de

Problems firing up	
Stove does not ignite	
Causes:	
Stove is not switched on	Switch on stove
Electric cable is not connected	Connect stove
Faulty ignition	Replace ignition
Operational problems	
Fan speed is constantly too high	
Causes:	
Flue pipes dirty	Clean and service them
Airflow sensor correct?	Check direction of the airflow sensor
Stove goes out after firing up	
Causes:	
Pellet quality OK?	Check pellet quality
Burner seat is incorrect	Control the burner
Chimney has too little draught	
Stove door is open	Close stove door
Faulty air sensor?	Check air sensor
Stove only runs at heating level 1	
Causes:	
Faulty room thermostat?	Connect room thermostat
Room is too warm	Wait for colder period
Stove goes out	
Causes:	
Pellet hopper is empty	Fill pellet hopper according to manufacturer's instructions
Viewing window is very black	
Causes:	
Is burner seat correct?	Check seat, clean holder
Unsuitable pellets	Check pellet quality
Level of ashes too high?	Clean burner pot and ash pans
Chimney does not draw air	Widen chimney
Too much soot in chimney	Check chimney, connection pieces
Window air washer not effective	Check and clean air washer
Pellets are not loaded	
Causes:	
Auger is still empty	Run the auger at maximum capacity according to manufacturer's instructions
Auger is blocked	Clean hopper and auger with vacuum cleaner
Faulty auger motor?	Contact service technician

5.3.4.6 SAFETY PRECAUTIONS FOR WORKING WITH PELLET STOVES
Pellet systems include components such as chains, springs and augers that are driven with considerable torques. This means that there is always a considerable risk of fingers becoming trapped when carrying out repair and assembly work. Therefore the power outlets should always be removed from the mains, and mechanical parts should not be under tension.

Standard wood pellets must be capable of igniting in any functioning pellet stove. Under no circumstances should damp or otherwise faulty pellets be ignited by mixing them with flammable liquids.

5.3.5 Central heating cookers
Wood logs can be used not just for heating but also in ranges for cooking and baking. Traditional wood-fired or coal-fired kitchen ranges are these days provided by some manufacturers with an additional heating function as central heating cookers (Figure 5.27).

Figure 5.27.
Modern central heating cooker
Photo: www.wamsler-hkt.de

Modern central heating cookers are used not just for cooking, baking and heating kitchens; they can heat the entire building, including the domestic hot water.

Technically this is achieved by surrounding part of the firebox with a water jacket, and by including other heat exchangers in the hot gas tubes that are coupled with the house's central heating system.

The surplus heat that is not used for cooking and baking is used for the central heating system or is stored as hot water in a storage tank (accumulator) integrated in the system.

Central heating cookers achieve an overall efficiency of more than 65%, because the heat radiated in the room where the range is installed is used for heating purposes and is not deemed to be a loss.

5.3.5.1 TECHNICAL SPECIFICATIONS FOR CENTRAL HEATING COOKERS
See Table 5.16.

Table 5.16.
Technical specifications for central heating cookers
Data: www.sesolutions.de

Primary application	Cooking, space heating
Installation location	Kitchen
Heating area	Kitchen, living space
Heat emission through radiation	Yes
Heat emission through convection	Yes
Heat emission through water heat exchanger	Yes
Type of combustion	Room-air-dependent
Combustion chamber	Closed
Output range (heat)	11–27 kW
Efficiency	> 65%
Usable fuel	Wood logs, wood briquettes
Ignition process	Manual, automatic
Flue temperature	250–320°C
Joint use of chimney with other systems	Yes
Required chimney diameter	To be calculated individually
Chimney soot fire resistance	1200°C
Moisture-resistant chimney	Yes
Prefabricated heating system	Yes
Fresh air demand	4 m^3 space volume/ kW
Safety clearance to combustible material at the front	> 0.80 m
Safety clearance to combustible material at the sides	> 0.20 m
Safety clearance to combustible material above	> 0.70 m
Safety clearance to combustible material on the floor	Fireproof base

5.3.5.2 STRUCTURAL REQUIREMENTS FOR CENTRAL HEATING COOKERS

Central heating cookers produce heat when operated. This is released as radiation through the window, or increases the temperature of the outer casing. For this reason, such stoves must always be kept at a safe distance from combustible building materials, furniture, curtains or other decorative fabrics. Manufacturers provide special spacers for this to enable the stoves to be integrated into modern fitted kitchens. When designing a kitchen with a central heating cooker, the following minimum clearances and provisions must be observed under all circumstances:

- There must be a minimum clearance of 70 cm to the oven glass door of the central heating cooker. In all cases, the individual requirements of the stove manufacturer must be observed.
- There is less risk of surrounding materials catching fire with the other areas of the stove such as the rear wall, side cladding and stovepipe. Therefore the distance to inflammable objects can be reduced to 20 cm from the stove. However, the individual specifications of the manufacturer are also decisive here.
- Kitchen cabinets above the central heating cookers must be placed with a minimum separation of 70 cm to the stove so as to prevent heat damage to the cupboard material or combustion of the contents.

The cooker should not be placed directly on combustible floors such as carpets, parquet or cork. It always requires a base made of a non-combustible material such as ceramic tiles, stone, glass or steel. Generally this base should extend 50 cm beyond the front of the stove and 30 cm on either side to ensure optimum fire protection. The individual requirements of the manufacturer are also decisive here when designing the fireproof base.

The most important elements of a central heating cooker are the air inlet and circulation openings. The air must be able to pass through these at all times. These openings must be kept open, otherwise poor combustion could lead to carbon monoxide poising, or even the risk of explosion.

Because they are heated up for cooking and baking, central heating cookers give off fluctuating amounts of heat. To be able to make optimum use of this heat in buildings, it is therefore advisable to integrate a storage tank (accumulator) in the building's central heating system (Figure 5.28).

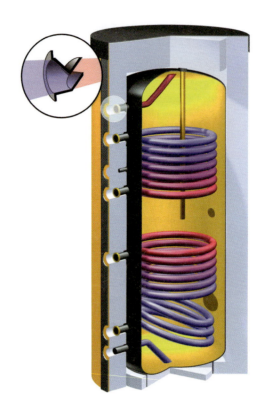

*Figure 5.28.
Storage tank with heat exchanger coils
Graphic: www.wagner-solartechnik.de*

Even if some manufacturers consider a storage tank of 25 l/kW to be sufficient, in practice storage tank volumes should be between 50 and 74 l/kW of thermal boiler capacity. If incorporating a solar thermal system, it is generally better to choose a larger storage volume.

The storage tank is installed between the flow and return pipes of the central heating cooker. Modern storage tanks can be connected to various space heaters at the same time, such as wood-fired heaters and solar thermal installations. The storage tank can also be used for supplying both space heating and domestic hot water.

5.3.5.2.1 FRESH AIR SUPPLY

Central heating cookers are classed as room-air-dependent hearth appliances. This also applies if the cooker is equipped with a separate external air inlet. There must be a balanced ratio between the room size and the output of the central heating cooker. Generally a room volume of at least 4 m^3 is required per kW of heat output from the cooker.

In addition, the safety limit for the maximum negative pressure differential is 4 Pa. For this reason it is expressly forbidden to operate devices that are capable of creating higher negative pressures in the room in which the stove is installed.

In particular, these devices include extraction hoods and ventilation systems. Technical measures should be used to ensure that no parallel operation is possible at any time, such as mutual locking of the circuits for the central heating cooker and extraction hood, or by ensuring that the latter is switched to recirculation mode. Alternatively, a sufficiently large air inlet opening should be provided in the affected room, the effectiveness of which must be proven through measurements.

5.3.5.2.2 FLUE GASES

For removing flue gases from central heating cookers, it is necessary to have a chimney that is approved for heating solid fuels. This must be able to withstand temperatures up to 400°C and have a soot fire resistance of 1200°C.

Flue pipes permitted for gaseous and liquid fuels may not be used for solid-fuel-fired central heating cookers. The flue pipe to the chimney should be short and sealed. Horizontal sections are generally permitted up to 0.5 m, but must have an upward inclination of >15°.

Only components produced by the cooker manufacturer should be used for connecting to the chimney, to ensure a good airflow and an airtight seal to the chimney.

5.3.5.2.3 MAINTENANCE

The grate of the central heating cooker should be checked daily for slag and clinkers; if necessary, this should be removed using the ash rake. Accumulated foreign bodies such as nails or stones, which can collect on the cooker grate through the use of building timber, should also be removed.

The air openings should also be cleaned if necessary. It is important here to prevent any ash from being pushed back through the grate as this could lead to blockages. For safety reasons, the ash should only be removed from central heating cookers once they have cooled down, to prevent waste material from being ignited by the hot ash.

5.3.5.3 TIPS FOR REMEDYING THE MOST FREQUENT OPERATING FAILURES OF CENTRAL HEATING COOKERS

See Table 5.17.

Table 5.17.
Most frequent operating problems with central heating cookers
Data: www.sesolutions.de

Problems firing up	
Wood does not ignite	
Causes:	
Wood is too thick	Chop the wood logs into smaller pieces
Wood is too damp	Dry out the wood for at least 1 year
Too little air supply	Open air regulating damper, control room air supply
Smokes escapes from the oven	
Causes:	
Ash door is briefly open	Ash door should be kept closed
Fuel is too damp	Dry out the wood for at least 1 year
Stove pipes are dirty	Thoroughly clean stove pipes
Operational problems	
Wood does not burn properly	
Causes:	
Flue draught is too weak	Install draught regulator
Cleaning doors are open	Firmly close cleaning doors
Too little fresh air supply	Fresh air duct must be kept free of blockages
Chimney is not sealed	Seal flue pipe and chimney
Chimney is overloaded	Size chimney
Temperature is too low when cooking or frying	
Causes:	
Oven door is briefly open	Keep oven door closed
Ash door is briefly open	Keep ash door closed
Too little wood	Put more wood in the stove
Grate sticks when shaken	
Causes:	
Grate is slagged up	Remove clinkers
Foreign bodies or nails	Remove foreign bodies or nails

5.3.6 Tiled stoves

Tiled stoves are fixed heating appliances constructed by skilled tradesmen. They are clad in stove tiles made of pure, dead-burned fireclay. Such heat-storage stove systems are large in size.

The tiled stoves, which are mostly situated centrally within buildings, heat areas that normally extend over several rooms. They are generally installed as additional heating to the existing central heating, and are often coupled to the central heating circuit. In general a distinction is made between basic and warm air tiled stoves.

5.3.6.1 BASIC TILED STOVES

The basic tiled stove (Figure 5.29) is a so-called heat-storage stove. Fireclay material and tiles directly surround its firebox. The heat that is generated there is emitted to the heat-storing ceramic cladding, from where it then gradually reaches the tile surface. From here it is radiated into the room space to provide heating.

Because of the slowness in emitting heat, a basic tiled stove requires at least 2 hours before reaching its full heat output. To shorten this time it is possible to build in heat exchanger tubes or ducts above the boiler. These are open on one or two sides, or can be closed with small doors. By opening the doors, they act as rapid heating surfaces and radiate their heat directly after being heated up. Food can be kept warm in the heat exchange ducts and, depending on the construction, it is also possible to bake foodstuffs.

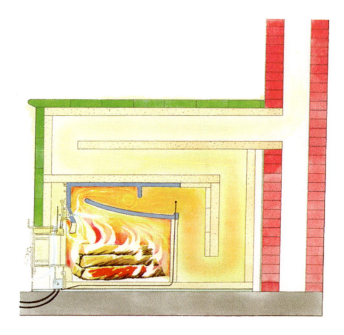

Figure 5.29.
Cross-section through a
basic tiled stove
Graphic: www.brunner.de

5.3.6.1.1 CONSTRUCTION OF A TILED STOVE

Figures 5.30–5.35 document the construction of an artistically designed tiled stove by a specialist company.

First of all the heat insert of the tiled stoves is mounted firmly on fireproof supporting blocks (Figure 5.30). The fireclay lining can be easily seen in the open heat

Figure 5.30.
Installing the heat insert
Photo: www.pernet.ch

insert. The electronics for controlling the heating insert can be seen on the stove door in the centre of the picture. To the right, the front cover for the insert is leaning against the wall.

Next the hot water pipes are installed, and the fireproof foundations are laid for the stove's ceramic heat exchanger (Figure 5.31). The hot flue gases are drawn through these and transfer their heat to the stove material.

Figure 5.31.
Laying the base
Photo: www.pernet.ch

During this phase of the stove construction the foundations are laid (Figure 5.32), and the first part of the ceramic cover is fitted. The stove's ceramic heat exchanger can be easily seen in the middle of the picture as a horizontal unit.

Figure 5.32.
Foundations for the heat exchanger
Photo: www.pernet.ch

In Figure 5.33 the form of the finished stove can already be visualised. In this phase the ceramic covering for the stove's seating area is already completed. The ceramic heat exchanger with its two inspection openings is clearly visible beneath the surface of the stove.

In the background is the extension of the heat exchanger to the chimney flue. The heating insert can also be directly connected to the chimney flue by opening a damper.

*Figure 5.33.
Formwork of the tile stove
Photo: www.pernet.ch*

The plastered formwork of the stove (Figure 5.34) gives a good impression of the stove's finished design. The ornamental metal cover for the heating insert stands out against the grey stove plaster.

*Figure 5.34.
Base plaster for the tiled stove
Photo: www.pernet.ch*

*Figure 5.35.
Finished tiled stove with creative design
Photo: www.pernet.ch*

In Figure 5.35 the creative design of the completed tiled stove can be admired in its finished form. Generally, there are hardly any limits to the design of tiled stoves, or to their architectural integration into buildings.

The advantage of basic tile stoves is their considerable ability to store heat. Once heated they can radiate an almost constant heat in the room for up to 24 hours. The disadvantage is the considerable slowness of the basic tiled stove system. A basic tiled stove can only be expected to give off heat after a heating-up phase of up to 2 hours.

5.3.6.2 WARM AIR TILED STOVES

Warm air tiled stoves also have an industrially produced, cast iron heating insert that is situated in the middle of a hollow space (heating chamber) (Figure 5.36). In operation, cool room air is sucked into the stove from below. This flows upwards through the heating insert and exits via the air grate and ducts as warm air. The heat of the stove is also radiated via the tile cladding. This type of stove emits heat more dynamically than basic tiled stoves.

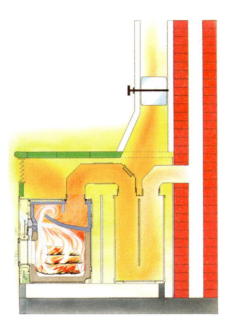

*Figure 5.36.
Cross-section through a warm
air tiled stove
Graphic: www.brunner.de*

With warm air tiled stoves the combustion and the tile walls are always separated. The heat generated in the firebox of the basic tiled stove and in the heating insert in the warm air tiled stove is absorbed by the tiles, directed to the surface and then radiated outwards or transferred to room air flowing past. The surface of the tiles is heated to an average temperature of between 50°C and 90°C.

In this temperature range the proportions of radiation and convection substantially correspond to the conditions that people perceive as comfortable. The tiles' low thermal conductivity means it is possible to touch them without burning oneself, even though they often have a surface temperature of 140°C near the firebox.

Around 60% of the heat created is emitted from the heating insert to the surrounding tile jacket, which in turn stores the heat and, on reaching the radiation temperature, radiates it into the room. The remaining 40% is emitted by convection of the air jacket.

The heating inserts for tiled stoves (Figure 5.37) have a wrought iron housing with a ceramic lining of fireclay material, in which wood logs or timber briquettes are burned. Depending on individual taste, there are heating inserts with large or small viewing windows made of quartz glass.

Modern heating inserts generally have a water heat exchanger for integrating the tiled stove with the central heating system. In addition, some manufacturers offer heating inserts with integrated storage tanks for hot water.

*Figure 5.37.
Heat inserts for tiled stoves
Photo: www.brunner.de*

The room air is drawn through the heating insert, entering below through the heating chamber. As it is drawn in through this passage the air rises. It then flows through air vents in the upper part of the tile wall and back into the room as warm air. The air circulation occurs completely by gravity.

Tiled stoves emit heat for a period of 6–24 hours. By stoking them once or twice a day, it is possible to provide continuous heat emission via the storage mass provided by the tiles. Modern tiled stoves have heating efficiencies between 75% and 89%.

The advantage of warm air tiled stoves is that they start to emit heat (via convection) before the tiles have heated up to the radiation temperature. A possible disadvantage is the very large proportion of heat emitted by convection, which could be found to be uncomfortable. Therefore, for reasons of comfort, it should be ensured that there is a very large tiled area (radiation area).

5.3.6.3 TECHNICAL SPECIFICATIONS FOR TILED STOVES
See Table 5.18.

Table 5.18. Technical specifications for tiled stoves
Data: www.sesolutions.de

Primary application	Space heating
Installation location	Living room
Heating area	Adjoining rooms
Heat emission through radiation	Yes
Heat emission through convection	Yes
Heat emission through water heat exchanger	Yes
Type of combustion	Room-air-dependent
Combustion chamber	Closed
Output range (heat)	4–15 kW
Efficiency	< 90%
Usable fuel	Wood logs, wood briquettes
Ignition process	Manual
Flue temperature	< 300°C
Joint use of chimney with other systems	Yes
Required chimney diameter	To be calculated individually
Chimney soot fire resistance	1200°C
Moisture-resistant chimney	Yes
Prefabricated heating system	No
Fresh air demand	4 m^3 space volume/ kW
Safety clearance to combustible material at the front	> 0.20 m
Safety clearance to combustible material at the sides	> 0.20 m
Safety clearance to combustible material above	> 0.70 m
Safety clearance to combustible material on the floor	Fireproof base

5.3.6.4 STRUCTURAL REQUIREMENTS FOR TILED STOVES
In contrast to wood-fired stoves, tiled stoves can be connected to chimneys that are already being used for other hearth appliances. However, a certified chimney sweep should always be consulted before connecting a hearth appliance to a chimney.

It is important to ensure that the tiled stove, flue pipe and chimney match one another in terms of performance. This task should also be left to specialists (tiled stove builder and/or certified chimney sweep).

Various types of wall material are used for tiled stoves. Apart from tiles, it is also possible to use soapstone or plaster. These days tiled stoves are very much considered to be design objects. Tiled stoves must be carefully and professionally designed, both in new and old buildings.

To achieve the pleasant 'mild radiant heat', it should be ensured that the radiant heat emitted by the tiles comes into contact with easily warmed surrounding surfaces such as walls and ceilings. It is therefore not very beneficial if the long wall of the tiled stove is placed so that it is directly opposite a window frontage. It is better to construct such a stove in the centre of a building so that the heat emitted can supply large parts of the house.

Tile stoves can be freestanding, or situated along a wall (if possible facing inwards), or in a corner. It is also possible to design them so that the outer surfaces of the stove extend into several rooms. When situated in only one room, there is greater heat efficiency if the distance from the tiled stove to the wall surface is at least 15 cm.

Every hearth appliance, whether freestanding or built-in, requires a load-bearing fireproof base. If tiled stoves are taken into account in time when designing new buildings, it is possible to provide any necessary strengthening. Existing ceilings, however, must be carefully examined to see how the beams or supports run and whether the ceiling construction fulfils the fire protection requirements.

The fireclay material installed in the stove construction consists mainly of silicic acid and clay substances. The panels or blocks that are used for the inner lining of tiled stoves must be fireproof and able to withstand temperatures of around 1200°C. Loam may be used as a binding material but not as a construction material as is the case with fireclay. Bricks are used only for building up plinths, providing footings, and laying foundations.

5.3.6.5 OPERATION OF TILED STOVES

Tile stoves are generally used as auxiliary heating. They are used in the transitional periods before the main heating season so that full heating does not have to begin until winter. Tiled stoves can also be used, however, in low-energy homes as the main form of heating. It is also possible to combine a central heating system coupled to a tiled stove with a solar collector for supplying domestic hot water.

Basic tiled stoves are suitable for heating one room or, when built centrally within a building, for heating several spaces. Warm air tiled stoves can heat up to three rooms. Here, the heated air can heat other rooms on the same floor or on the floor above by using shafts or ducts.

The tile stove's size and type of construction depend on the heat demand of the room being heated. When determining the tiled stove output it should also be considered whether it is to be used for full heating or auxiliary heating.

It should also be discussed with the tiled stove builder as to which form of tiled stove is suitable for the client's particular needs. Depending on the type of construction, the heat output from tiled stoves ranges between 600 W and 1000 W per square metre of tiled stove area. Warm air tiled stoves heat the room air immediately on being heated, but have less storage capacity. Basic stoves are slower, but emit a characteristic heat that is very pleasant.

Apart from correctly igniting the stove, the most important factor when heating with tiled stoves is the correct use of the damper. The damper enables the hot flue gases to be conducted directly into the chimney, allowing it to be heated more quickly and therefore creating a better draught.

5.3.6.6 TIPS FOR REMEDYING THE MOST FREQUENT OPERATING FAILURES OF TILED STOVES

See Table 5.19.

Table 5.19.
Most frequent operating problems with tiled stoves
Data: www.sesolutions.de

Problems firing up	
Wood does not ignite	
Causes:	
Wood is too thick	Chop the wood logs into smaller pieces
Wood is too damp	Dry out the wood for at least 1 year
Too little air supply	Open air regulating damper, control room air supply
Operational problems	
Wood burns without bright yellow flame, smoulders and even goes out	
Causes:	
Flue draught is too weak	Install draught regulator, clean chimney connection
Wood is too damp	Dry out the wood for at least 1 year
Too little air supply	Open air regulating damper, control room air supply
Too much soot forms, the stove walls do nor remain clean	
Causes:	
Wood is too damp	Dry out the wood for at least 1year
Too little air supply	Open air regulating damper, control room air supply
Too little wood	Put more wood in the stove

Table 5.19. continued

Although the fire burns strongly, the stove does not become warm	
Causes:	
Flue pipe to chimney is open	Close damper
Flue draught is too strong	Install draught regulator
Wood burns too quickly with low heat output	
Causes:	
Flue draught is too strong	Install draught regulator
Wood is chopped into pieces that are too small	Insert larger pieces
Stove operation is incorrect	Read instructions for use
Smoke escapes through the stove door	
Causes:	
Stove door is not closed	Make sure front door is closed tightly
Too little air supply	Open air regulating damper, control room air supply
Chimney cross-section is too narrow	Widen chimney
Too much soot in chimney	Contact chimney sweep
Wind-induced downdraught from chimney	Install chimney cap
The chimney becomes wet and sooty, condensation drips down from the chimney	
Causes:	
Wood is too damp	Dry out the wood for at least 1 year
Flue gas is too cold	Heat up chimney with maximum output
Chimney cross-section is too large	Refurbish chimney

5.3.7 Log-fired central heating boilers

Log-fired central heating boilers (Figure 5.38) are conventional heating systems heated with chopped wood or briquettes; they are located in a separate boiler room, and can cover the complete heat demand of an entire building. These wood-gasifying boilers, as they are also known, are therefore used to provide the complete heat supply for single dwellings and large buildings.

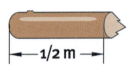
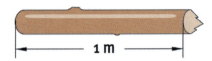

Figure 5.38.
Modern log-fired central heating boilers
Photo: www.koeb-schaefer.com

Wood logs 25–100 cm long are used as the fuel; they are stacked in the boiler on a firebed. The resulting low-temperature fire produces wood gases, which are drawn into a secondary fire chamber where they are burnt completely. See Figure 5.39.

Log-burning boilers clearly separate the two different types of combustion. See Figure 5.40. In the primary combustion chamber the wood is heated and, as a result of burning with oxygen, inflammable wood gases are released. These are drawn through an induced-draught fan into a combustion chamber lined with ceramic material or high-temperature-resistant steel. This utilizes the wood gases created in the first boiler combustion chamber. The wood gas is mixed with the necessary secondary oxygen in a whirl chamber and burnt while being drawn through a combustion plate. Figure 5.41 shows the wood gas flame of a wood-gasifying boiler.

SMALL COMBUSTION SYSTEMS 153

*Figure 5.39.
Cross-section through a log-fired
central heating boiler
Graphic: www.koeb-schaefer.com*

*Figure 5.40.
Cut-away view and photo of a log-fired
central heating boiler
Graphic/photo: www.guntamatic.at*

The hot flue gases from both combustion stages are drawn through heat exchangers integrated in the boiler, where they transfer their heat to circulating water in the heating system. Thus the flue gases are cooled to temperatures below 200°C before being drawn through the chimney to the outside.

 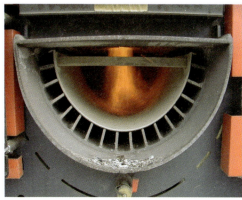

*Figure 5.41.
Burner plate and wood gas flame
Graphic/photo: www.froeling.at*

The use of a fan enables the combustion to be kept substantially independent of the surrounding conditions. This means that the draught conditions in the chimney play a less significant role.

Fans enable the primary and secondary combustion stages in modern log-fired boilers to be precisely coordinated with each other. In addition, they also enable a larger pressure loss to be overcome in the firebox, which is required to achieve an optimal mixture of secondary air and boiler-created flammable gases in the secondary combustion chamber.

There are log-fired central heating systems with two different control strategies on the market. Thermostatically controlled systems adjust the heat produced in the boiler to the boiler water temperature and the demand on the domestic heating system. Other boilers monitor the oxygen content in the boiler's flue gas with a lambda sensor and can therefore ensure that there is always optimum combustion. Such technical advances also enable log boilers to attain an efficiency of more than 90%.

By throttling back the fan, it is also possible for central wood-gasifying boilers to burn partial loads up to 50% of the nominal load. Despite the high-tech control technology, this nevertheless lowers the efficiency of the boiler under such circumstances. Therefore it is sensible to integrate an accumulator tank that balances out the fluctuations between the heat demand and the supply. This addition also allows central log-fired boilers to be optimally combined with solar thermal installations.

In contrast to single hearth appliances or space heaters in rooms, log-fired boilers are installed in separate boiler rooms with a connection to the central heating system. Their means of construction is therefore fundamentally different, as log-fired boilers are designed to prevent heat from being emitted via the boiler sides to the surrounding space. Generally these boilers are well insulated and look more similar to oil or gas-fired boilers than to traditional wood-burning hearth appliances.

Figure 5.42.
Air circulation in a log-fired central heating boiler
Graphic: www.guntamatic.at

5.3.7.1 TECHNICAL SPECIFICATIONS FOR WOOD-GASIFYING BOILERS
See Table 5.20.

Table 5.20.
Specifications for wood-gasifying boilers
Data: www.sesolutions.de

Primary application	Space heating, hot water
Installation location	Boiler room
Heating area	Entire building
Heat emission through radiation	No
Heat emission through convection	No
Heat emission through water heat exchanger	Yes
Type of combustion	Room-air-dependent
Combustion chamber	Closed
Output range (heat)	5–150 kW
Efficiency	< 90%
Usable fuel	Wood logs, wood briquettes
Ignition process	Manual, automatic
Flue temperature	150–200°C
Joint use of chimney with other systems	Yes
Required chimney diameter	Calculate individually
Chimney soot fire resistance	1200°C
Moisture-resistant chimney	Yes
Prefabricated heating system	Yes
Fresh air demand	Calculate individually
Safety clearance to combustible material at the front	> 0.80 m
Safety clearance to combustible material at the sides	> 0.50 m
Safety clearance to combustible material above	> 0.70 m
Safety clearance to combustible material on the floor	Fireproof base

5.3.7.2 STRUCTURAL REQUIREMENTS FOR WOOD-GASIFYING BOILERS

Wood boilers with an overall heat output of more than 50 kW are normally only allowed to be installed in their own boiler room. These should have a minimum ceiling height of 2 m and a minimum room volume of 8 m^3.

It is also important that the boiler's air feed is substantially dust-free and free of halogenated hydrocarbons such as are produced by sprays, pigments, paints and solvents. These can impair the operation of the boiler.

The internal temperature of the boiler room should not exceed 40°C even when the boiler is operating. It is also not permitted to store combustible materials, fluids or gases in the boiler room. As a consequence, no more than a day's supply of firewood may be stored beside the log-burning boiler.

The minimum clearance of the boiler to the wall is generally 50 cm, whereby the front door of the boiler should be at least 80 cm from the next wall. The boiler must have a clearance of at least 1 m from where the daily supply of firewood is stored.

The ash that is produced when operating the boiler must be stored in non-combustible containers with lids. Boiler room sizes of 20 m^2 require fire extinguishers containing a minimum of 6 kg of powder. Fire extinguishers with 12 kg of powder are required for boiler rooms between 20 m^2 and 50 m^2.

It is advisable to install a storage tank for wood-gasifying boilers in order to be able to cover the daily fluctuations in heat demand. The storage tank should also be able to absorb the resulting heat from a loaded boiler to ensure the optimum use of the fuel.

Even if some manufacturers consider a storage tank of 25 l/kW to be sufficient, in practice storage tank volumes should have between 50 l/kW and 74 l/kW of thermal boiler capacity. If incorporating a solar thermal system, it is generally better to choose a larger storage volume.

Optimum boiler efficiencies are reached when the temperature of the central heating system is between 70°C and 85°C. The temperature in the central heating system should be at least 60°C, otherwise most control systems cannot be operated properly.

5.3.7.2.1 FRESH AIR SUPPLY

For the boiler to operate safely, it must have an optimum supply of incoming and exhaust air. The negative pressure in the boiler room must not exceed 4 Pa. This is achieved by providing supply openings with a free cross-section of at least 300 cm² for all boiler sizes up to 50 kW. Boilers with a greater output require another 2.5 cm² of supply area in addition to this value for every additional kW.

The supply and exhaust ducts should, as far as possible, be arranged opposite each other to achieve good thermal suction. The openings must always be covered with a grille to prevent foreign bodies such as leaves or small animals from entering the boiler room.

Here grilles with a maximum mesh spacing of 10 mm have proven reliable. Modern chimneys often already have rear ventilation that can replace the exhaust opening.

5.3.7.2.2 FLUE GASES

For removing flue gases from the log-burning boiler, it is necessary to have a chimney that is approved for heating solid fuels. This must have a soot fire resistance of 1200°C. Note that the flue temperatures in the chimney can sink below 160°C and sometimes even below 90°C, making it possible for condensation to occur.

For this reason, the chimney should be constructed of moisture-proof material and be highly insulated. This applies particularly to the passage of the chimney through the attic of a house. In addition, the dew point should also be calculated. Most boiler manufacturers require the chimney to have a minimum diameter of 16 cm.

Flue pipes permitted for gaseous and liquid fuels may not be used for wood-gasifying boilers. If the chimney has a draught that is greater than 20 Pa, a draught controller should be used.

The flue pipe to the chimney should be short (length < 2.0 m) and airtight. The connection to the chimney must always be laid so that it slopes upwards (>15°). In practice, upward tilts of 30–45° in the direction of flow have proven reliable. Furthermore, the connection should be insulated (> 5 cm heat resistant) and, if possible, executed without any bends. The entry to the chimney should facilitate the flow, and should curve upwards.

When wood-gasifying boilers are first fired up, a certain amount of positive pressure is to be expected. For this reason, the flue pipe to the chimney must be laid so that it is completely airtight. Here it is possible to use temperature-resistant silicone as sealing compound; alternatively, the flue pipe can be welded tight. Furthermore, it is sensible to lay a flexible and insulated flue pipe to the chimney to improve the noise insulation.

5.3.7.3 NOTES ON OPERATING WOOD-GASIFYING BOILERS

5.3.7.3.1 MAINTAINING THE BOILER

Before heating up each time, excess ash should be removed from the log-fired boiler's loading chamber. Here the turbo or combustion plates of the wood-gasifying boiler should be lifted up and any existing ash swept into the ash pan with a brush or glove. The secondary air openings in the whirl chamber should also be kept free.

Under no circumstances should ash particles be forced through the burner hole, as this could damage the burner nozzles. Leaving some ash in the combustion chamber, however, improves ignition in the boiler because it closes the gap between the turbo or combustion plate and the boiler floor, and prevents unwanted air from penetrating.

The ash pan can then be removed and the ash disposed of. Every week the sides of the ash pan should be cleaned with an appropriate cleaning tool.

Once a month the space between the upper and lower plate must be cleaned. The turbo or combustion plates are removed and the entire ash is taken from the combustion chamber. In addition, all accessible combustion chamber parts in the boiler should be dismantled and cleaned.

Once every three months it is necessary to clean the fan to maintain the boiler output and draught. The cover parts are dismantled, and the deposits on the fan blades are removed with a brush.

5.3.7.3.2 SERVICING
Wood-gasifying boilers must be serviced once a year. Manufacturers normally offer a comprehensive maintenance contract, which provides complete servicing of the boiler components and the entire technical equipment. Often, as part of this servicing, switches and ignition elements are replaced or checked to see that they function correctly.

Apart from maintaining the function of the devices, it is also essential that boilers are serviced for safety reasons. Only boilers that are regularly serviced can be safely operated in the boiler room.

5.3.7.4 TIPS FOR REMEDYING THE MOST FREQUENT OPERATING FAILURES OF CENTRAL WOOD-GASIFYING BOILERS
See Table 5.21.

Table 5.21. Most frequent operating problems with wood-gasifying boilers. Data: www.sesolutions.de

Problems firing up	
Wood does not ignite	
Causes:	
Wood is too thick	Chop the wood logs into smaller pieces
Wood is too damp	Dry out the wood for at least 1 year
Too little air supply	Blower is not running, switch on boiler
Smoke comes out of the stove door	
Causes:	
Filling door or the ash door is open	Close doors
Flue draught is too little	Draught regulator is set too low
Flue draught is too little	Chimney is still cold
Flue draught is too little	Flue pipe is not sealed or has too many bends
There are no wood gas flames, only smoke forms	
Causes:	
Fill door lock is not engaged	Close door, blower switches on
Wood is too thick	Chop the wood logs into smaller pieces
Wood is too damp	Dry out the wood for at least 1year
Firebed is too small	Flue pipe is not sealed or has too many bends
Ash is too deep	Clean burner plate or grate
Fire does not get any air	Blower is stuck
No boiler draught	Upper burner plate is not sealed
No boiler draught	Lever for start-up flap is loose
Fan does not work	
Causes:	
Start button has not been pressed	Press start button
Boiler is not switched on	Switch on boiler
Boiler is without electricity	Connect plug
Operational problems	
Wood gas flame is too small boiler has no output	
Causes:	
Wood is too thick	Chop the wood logs into smaller pieces
Wood is too damp	Dry out the wood for at least 1year
Hollow burning in the boiler	Rattle wood
Burner plate is dirty	Clean burner plate
Fan is dirty	Clean fan
Fan is turned down	Turn it up
Heating circuit is too warm	Wait until heating circuit has discharged heat
Flames appear from the open ash door the boiler has no output	
Causes:	
Flue passages are blocked	Clean boiler
Flue pipe is dirty	Clean flue pipe connected to chimney
Chimney is filled with soot	Have chimney cleaned
No flue draught	Have chimney checked
Flame is very large and appears from the open ash door	
Causes:	
The wood is very dry	Use wood with approximately 18% moisture content
The wood pieces are too small	Load large pieces
Intermediate spaces in boiler are full	Clean boiler
Fan output is too high	Throttle back fan

Table 5.21. continued

Boiler heats too frequently above 95°C flame is normal Causes:	
Boiler output is too large	Permanently throttle back boiler output
Storage tank is too small for the system	Integrate larger storage tank
Circulation pump output is too small	Replace circulation pumps
Too much wood	Load less wood
Flame is dark yellow and flickers the fan runs Causes:	
Burner area is dirty	Clean burner area
Fan is dirty	Check fan output
Plastic in the fuel wood	Cool down boiler and clean it out
Boiler circuit pump switches on too frequently Causes:	
Boiler is rinsed through too strongly	Switch the boiler circuit pump to a lower level
Faulty boiler return temperature control	Check return temperature control
Boiler temperature is not reached	Heat for longer and more powerfully
Firebed is not maintained Causes:	
Wood is too damp	Set fan thermostat to higher temperature
Wood pieces are too small	Put large pieces in with next load
Flue draught too strong	Adjust draught regulator

5.3.8 Central pellet boilers

As well as the pellet stoves already described for living spaces, there are also pellet boilers fuelled by standard wood pellets, located centrally in buildings. As with wood-gasifying boilers, they are also installed in separate boiler rooms. They offer a complete alternative to heating with fossil fuels in all areas of heating, including spacing heating and domestic hot water provision.

Technically, there are three different ways of loading wood pellet boilers: bottom-feed, the retort system, and top-feed. The choice of which system to use depends on the boiler manufacturer. The principles of operation and the individual advantages and disadvantages are described below.

5.3.8.1 BOTTOM-FEED SYSTEM

A pellet boiler with bottom-feed loading consists of a drive system with electric motors and control systems outside the boiler and a burner component within the boiler. See Figure 5.43.

Figure 5.43.
Bottom-feed burner for a pellet boiler
Photo: www.paradigma.de

5.3.8.1.1 OPERATING PRINCIPLE

The wood pellets are loaded via a discharge auger (screw conveyor) into the combustion area. The primary air is drawn in via a ring-shaped steel grate similar to a car brake disc.

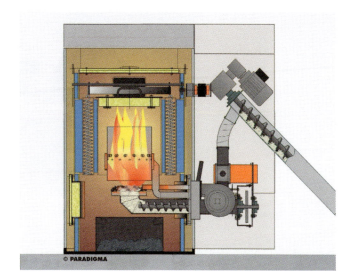

Figure 5.44.
Cross-sectional drawing of a pellet boiler with bottom-feed burner
Graphic: www.paradigma.de

Using secondary air holes or, as in Figure 5.44 using secondary air tubes, the post-combustion of the carbonization gas occurs in the heatproof combustion areas of the boiler, whereby the carbonization gas mixture is thoroughly mixed with secondary air. The hot flue gases that are created during the entire burnout process are then drawn through water-filled heat exchangers to enable the boiler to transfer heat to the central heating system.

5.3.8.1.2 TECHNICAL ADVANTAGES
It is easy to determine the pellet level in the burner pot, and this generally controls itself.

5.3.8.1.3 TECHNICAL DISADVANTAGES
The wood pellets come directly into contact with the fire zone. This means that there is a risk of burn-back into the hopper. This type of firing is technically sluggish because considerable afterheat is created in the burn plate, which is always full. The continual or intermittent movement of the auger can compact the fuel or destroy the pellets. It is also possible for an inhomogeneous firebed to form, allowing the pellets to end up in the ash areas without being burned.

Figure 5.45.
Bottom-feed burner in action
Photo: www.paradigma.de

5.3.8.2 RETORT SYSTEM
A pellet burner with retort firing (Figure 5.46) has a similar construction to that of a bottom-feed system. The conveyor mechanism and the control electronics are also situated outside the boiler, whereas the combustion zone is inside it.

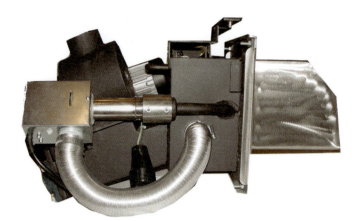

*Figure 5.46.
Retort burner for pellet boilers
Photo: www.gilles.at*

5.3.8.2.1 Operating principle

The wood pellets are loaded via an auger from the side into a steel container or a firebrick-lined combustion chamber.

The primary air is fed underneath the pellets. The secondary air is fed via a ring or pipe into the combustion zone and ensures the post-combustion and thorough mixing of the carbonization gas mixture with the post-combustion air. With retort burners the ash drops through the grate into an ash pan.

On route to the chimney the rising hot exhaust gases are drawn through a heat exchanger, which transfers the pellet heat into the central heating system of the house.

*Figure 5.47.
Cross-sectional drawing of a pellet
boiler with retort burner
Graphic: www.hargassner.at*

5.3.8.2.2 Technical advantages

The retort system uses a type of firing that develops little afterheat and therefore responds quickly when changes are made to the control settings.

5.3.8.2.3 Technical disadvantages

With retort firing, wood pellets come directly into contact with the fire zone. This means that there is a risk of burn-back. As with bottom-feed firing, the auger compacts the fuel, which could create an inhomogeneous firebed with poor combustion. With retort firing, more ash often drops down than with the other kinds of wood pellet firing system.

5.3.8.3 TOP-FEED SYSTEM

The top-feed system (Figure 5.48) uses a completely different conveyer philosophy for loading wood pellets into the burner. Here it is done by gravity. Inside the boiler the wood pellets are loaded from the hopper via an auger and then drop down through a pipe or chute into the burner pot.

Figure 5.48.
Top-feed burner for a pellet boiler
Photo: www.viessmann.de

With pellet boilers using top-feed systems (Figure 5.49), the fire is supplied with primary and secondary air directly in the burner pot. This results in complete burnout of the pellets and the flammable gases given off by them. The resulting flue gases are

Figure 5.49.
Cross-sectional drawing of a pellet
boiler with top-feed burner
Graphic: www.kuenzel.de

then fed upwards through heat exchangers within the boiler, which transfer the heat generated into the central heating system of the building.

The ash that is created in the burner pot also falls by means of gravity into an ash box or pan, from where it can be removed as part of the regular maintenance of the boiler.

5.3.8.3.1 Technical advantages

With top-feed systems the pellet conveyer mechanism is not directly connected to the fire zone, so there is no risk of burn-back into the storage hopper. In addition, this fuel loading method achieves a homogeneous, non-compacted firebed that burns with reduced ash fall. Pellet stoves with top-feed systems have a wear-resistant construction that can be equipped with an automatic cleaning system for the burner grate.

5.3.8.3.2 Technical disadvantages

It is difficult to monitor the pellet level with this type of construction. The monitoring must be done with a level gauge.

5.3.8.4 TECHNICAL SPECIFICATIONS FOR WOOD PELLET BOILERS
See Table 5.22.

Table 5.22. Technical specifications for wood pellet boilers
Data: www.sesolutions.de

Primary application	Space heating, hot water
Installation location	Boiler room
Heating area	Entire building
Heat emission through radiation	No
Heat emission through convection	No
Heat emission through water heat exchanger	Yes
Type of combustion	Room-air-dependent
Combustion chamber	Closed
Output range (heat)	5–35 kW
Efficiency	< 90%
Usable fuel	Wood pellets
Ignition process	Automatic
Flue temperature	150–200°C
Joint use of chimney with other systems	Yes
Required chimney diameter	Calculate individually
Chimney soot fire resistance	1200°C
Moisture-resistant chimney	Yes
Prefabricated heating system	No
Fresh air demand	Calculate individually
Safety clearance to combustible material at the front	> 0.80 m
Safety clearance to combustible material at the sides	> 0.50 m
Safety clearance to combustible material above	> 0.70 m
Safety clearance to combustible material on the floor	Fireproof base

Wood pellets are predominantly used for supplying heat in private residential buildings or small commercial buildings. They are used mainly in the small- and medium-output classes up to 50 kW. Here, automatically stoked wood pellet boilers provide a technically well-developed and economically viable alternative to oil and gas heating.

In contrast to oil or gas-fired heating systems, pellet-fired systems are operated with a batch control system. This means that pellet boilers constantly burn a small amount of pellets before the stoking mechanism feeds new fuel to the firing zone. As a result it is impossible to meter the heat demand precisely, so there is always a slight delay of about 10–15 minutes before the heat demanded is emitted.

A sudden heat demand, such as when showering, can lead to problems if the central heating system has insufficient hot water. This problem can be solved, however, with hot water storage tanks (accumulators). These lengthen the burn intervals, as the heating system can accept more heat. This increases the efficiency and reduces the combustion emissions.

In addition, storage tanks allow solar thermal systems or other heat generators to be incorporated into the central heating system, lowering the annual fuel demand

Figure 5.50.
Pellet boiler with connection pipes
Photo: www.gilles.at

Figure 5.51.
Pellet–solar thermal
combination heating
Graphic: www.wagner-solartechnik.de

Figure 5.52.
Storage tank
Photo: www.paradigma.de

(Figure 5.51). Thus a well-designed solar thermal installation with a sufficiently sized storage tank can cover the entire hot water demand of a house outside the heating periods.

It is particularly recommended to install hot water storage tanks in buildings with low heat demands. Generally, installing a storage tank increases the convenience of the system (Figure 5.52).

Depending on the central heating system used in the home, an accumulator size of at least 25 l/kW of heat output from the boiler is recommended. If it is intended to incorporate a solar thermal installation, accumulator sizes between 50 and 75 l/kW of pellet heat output are recommended.

Currently, pellet central heating systems with a heat output of around 5–50 kW are available. Pellet heating boilers provide an ideal solution for environmentally aware homeowners in single- and multiple-family dwellings that have a low-energy construction, a heat output of 10–40 kW, and an annual energy demand of around 2000–20,000 kWh.

Such pellet boilers require up to 5000 kg of pellets per year. These are generally delivered in tankers and can be stored in different types of storage room. When designing a pellet heating system, it is important that the following prerequisites are fulfilled on the transport route up to a distance of 30 m from the building.

- street width > 3.00 m;
- clearance height > 4.00 m.

Most pellet suppliers use a 30 m long air injection hose and a 4 m long suction hose for the injected air. The fan used for this requires an electricity source close to the building with 230 V and a 16 A fuse.

Pellet deliveries are an essential component when heating with wood pellets. Therefore the constructional requirements in terms of access routes must be clarified in advance and ensured on a long-term basis.

5.3.8.4.1 Fresh air supply

For the boiler to operate safely it musty have an optimum supply of incoming and exhaust air. The negative pressure in the boiler room must not be greater than 4 Pa. This is achieved by providing air supply and exhaust openings with an open cross-section of at least 150 cm^2 for all available pellet boiler sizes.

The supply and exhaust ducts should, as far as possible, be arranged opposite each other to achieve good thermal suction. The openings must always be covered with a grille to prevent foreign bodies such as leaves or small animals from entering the boiler room.

Here grilles with a maximum mesh spacing of 10 mm have proven reliable. Modern chimneys often already have rear ventilation that can replace the exhaust opening.

5.3.8.4.2 Flue gases

For removing flue gases from pellet boilers, it is necessary to have a chimney that is approved for heating solid fuels. This must have a soot fire resistance of 1200°C. Note also that the flue temperatures in the chimney can sink below 160°C and sometimes even below 90°C.

For this reason, the chimney should be constructed of moisture-proof material and be highly insulated. This applies particularly for the passage of the chimney through the attic of a house. Most boiler manufacturers require the chimney to have a minimum diameter of 14 cm.

Flue pipes permitted for gaseous and liquid fuels may not be used for pellet boilers. If the chimney has a draught that is greater than 20 Pa, a draught controller should be used. Pellet boilers require a minimum flue draught of 5 Pa.

The flue pipe to the chimney should be short (length < 2.0 m) and leakproof. The connection to the chimney must always be laid so that it slopes upwards (>15°). In practice, upward tilts of 30–45° in the direction of flow have proven to be reliable. Furthermore, the connection should be insulated (> 5 cm heat resistant) and, if possible, executed without any bends. The entry to the chimney should facilitate the flow, and should curve upwards.

With log boilers, the flue gas tubes in the boiler's heat exchanger often require considerable cleaning. Modern pellet boilers, by contrast, are easy to clean in this respect, possessing automatic cleaning devices for the flue gas tubes (Figure 5.53).

When firing up pellet boilers with a cold chimney, it is to be expected that there will be some positive pressure in the flue pipe. For this reason, the flue pipe connection to the chimney should be laid so that it is completely airtight. Here it is possible to use temperature-resistant silicone as sealing compound or, alternatively, the flue pipe can be welded tight. Furthermore, it is sensible to lay a flexible and insulated flue pipe to the chimney to improve the noise insulation.

A rule of thumb for measuring chimneys is that chimneys with diameters of 14 cm can be used with pellet boilers with outputs up to 15 kW. Chimneys with 16 cm diameters can be used with pellet boilers with outputs between 20 kW and 25 kW. With greater boiler outputs, chimney diameters of 18 cm are recommended.

Figure 5.53.
Heat exchanger for a pellet boiler
with a cleaning mechanism
Photo: www.paradigma.de

5.3.8.4.3 Operation

Pellet boilers are fully automatic heating systems. If the technical installation of the system conforms to the quality of the fuel, it is necessary only to remove the ash pan every 2–8 weeks. Most pellet boilers are equipped with an ash compression system to increase the convenience.

It is important that the system is switched off before removing the ash so that the ash can cool off in the ash pan. The handles for emptying the ash pan of a pellet boiler are shown in Figure 5.54 by way of example.

Normal pellet ash is grey-brown and gritty. If the boiler is not correctly set, the ash is jet black because charcoal has formed and the pellets have not been correctly burned. With incorrectly set pellet boilers with bottom-feed or retort burners, it is also possible for unburned pellets to fall into the ash pan. Should these technical faults be

Figure 5.54.
Cleaning pellet boilers
Photo: www.oekofen.at

repeatedly observed, the boiler manufacturer's customer repair service should be notified. This is due to a problem with the electronic boiler control or with the level monitoring.

Pellet boilers generally have sophisticated monitoring technology that can precisely pinpoint the malfunction. Whereas electronic faults cannot be fixed without the customer repair service, combustion problems can often be remedied on site.

Frequent causes of burner malfunctions include:

- no more pellets in the storeroom;
- level indicator in the intermediate hopper faulty;
- motor blockage in the auger;
- electric igniter faulty.

5.3.9 Woodchip boilers

In addition to wood pellets, woodchips can also be used as fuel in automatic boilers. Technically, automatically fed woodchip boilers are very similar to wood pellet boilers. The woodchips are generally fed into the boiler with spiral or screw augers.

As with wood pellets, woodchips are a bulk material. However, woodchips are far less homogeneous than pellets, owing to their means of production with mechanical chippers. As a result this increases the risk of arching and blockages of the conveyor systems and hoppers.

For this reason, woodchip boilers are more robust and larger than pellet boilers. They therefore have a minimum combustion capacity of 35 kW. Depending on the design, woodchip heating systems are also produced as large-scale plants that can generate a heat output of several megawatts. This section, however, will consider only woodchip boilers with small output ranges. Chapter 6 describes larger plants.

Figure 5.55.
Woodchip boiler with extraction device
Photo: www.hargassner.at

Figure 5.55 shows a woodchip boiler with an automatic bunker extraction device. The protruding rotor blades of the revolving spiral are used to loosen the woodchip structure, thus preventing the extraction point in the bunker from becoming blocked. The pressure plate fixed to the auger box relieves the pressure of the grain structure in the bunker. This prevents bridging in the woodchip structure and therefore stops the conveyor screw from turning around empty.

In the picture an ash pan with a red lid can be seen in front of the actual boiler, in which the ash from the combustion process is collected. Just as with pellet boilers, the fuel is ignited using a hot air gun. Its fan blower and red-coloured casing can be seen to the right.

The cross-section in Figure 5.56 shows how the woodchips are conveyed via the in-built augers to the combustion zone.

Although the technology of these boilers is well developed and efficient, woodchip boilers are not suitable for domestic homes because of the high basic costs of the boiler, storage bunker and conveyer equipment. However, in economic terms they are ideal for applications where wood pellets are no longer suitable – that is, for apartment and communal buildings.

*Figure 5.56.
Cross-section through a
woodchip boiler
Graphic: www.hargassner.at*

5.3.9.1 TECHNICAL SPECIFICATIONS FOR WOODCHIP BOILERS
See Table 5.22.

*Table 5.23.
Technical specifications for
woodchip boilers
Data: www.sesolutions.de*

Primary application	Space heating
Installation location	Boiler room
Heating area	Entire building, district heating
Heat emission through radiation	No
Heat emission through convection	No
Heat emission through water heat exchanger	Yes
Type of combustion	Room-air-dependent
Combustion chamber	Closed
Output range (heat)	35–7000 kW
Efficiency	< 90%
Usable fuel	Woodchips
Ignition process	Automatic
Flue temperature	150–200°C
Joint use of chimney with other systems	No
Required chimney diameter	Calculate individually
Chimney soot fire resistance	1200°C
Moisture-resistant chimney	Yes
Prefabricated heating system	No
Fresh air demand	Calculate individually
Safety clearance to combustible material at the front	> 0.80 m
Safety clearance to combustible material at the sides	> 0.50 m
Safety clearance to combustible material above	> 0.70 m
Safety clearance to combustible material on the floor	Fireproof base

5.3.9.2 STRUCTURAL REQUIREMENTS FOR WOODCHIP BOILERS
Woodchip boilers always require their own boiler rooms, as well as storage areas close by. It should be possible for the storage areas to be supplied easily with wood delivered by lorry, while being close enough to the boilers to allow the use of low-cost augers with a minimum conveyor length. Examples of such storage concepts are shown in the last part of this chapter.

Because woodchip boilers have a large range of outputs, from 35 kW to 7 MW, it is impossible to make general statements about the types of flue pipe, air supply duct or other technical elements. In the small output range, up to 100 kW, demands on the flue and air supply are identical to those for wood-gasifying boilers. The same applies for the hydraulic connection to the heating circuit.

5.3.9.3 OPERATION
With heat emissions optimally adjusted to the buildings, woodchip boilers are continually fired during the heating period. During this period they are not switched off. Instead their output is simply modulated according to need, using their induced-draught fans.

Programmable logic controls (so-called PLC systems) control the ignition and complete system operation. This enables such a system to continually adapt itself to meet the heat demand, ranging from a rated load of 100% to a partial load of 30%. The electronic combustion control ensures almost complete burnout, with a high efficiency of 87–90% of the input energy.

*Figure 5.57.
Automatic cleaning device
Photo: www.hargassner.at*

Modern woodchip boilers are designed for fully automatic operation (Figure 5.57). This does not normally require regular cleaning work of the boiler, grate or flue systems. If, however, an automatic ash removal system is dispensed with for reasons of cost, the accumulated ash should generally be removed once a week.

5.3.10 Combination boilers

Combination boilers (Figure 5.58) can be used for applications where a wide range of fuels is available, such as woodchips, wood logs or sawdust. They are similar in construction to wood-gasifying boilers.

In addition to a stoking hole for wood logs, these boilers also have a side inlet for woodchips. Their construction enables them to accept various fuel geometries. A wide variety of wood products can be fired in them, such as sawdust, pellets, woodchips, logs, unrefined waste wood and construction lumber.

*Figure 5.58.
Combination boiler for wood logs and woodchips
Graphic: www.koeb-schaefer.at*

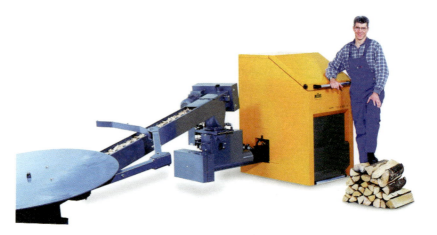

*Figure 5.59.
Combination boiler with automatic and manual loading
Photo: www.koeb-schaefer.at*

Combination boilers can prove to be particularly efficient at locations producing a sufficient amount of timber waste material and different grain sizes, such as in a joinery workshop. In such application areas they provide the best choice for a heating system. Combination boilers are also of interest for builders who desire particularly high flexibility in terms of the different fuels that can be used for supplying heat.

5.4 Basic design considerations

As was explained in the last section, there are a large variety of designs available for small combustion systems fuelled with wood. If these combustion systems are designed as single hearth appliances, you require no other construction components than a chimney. If the wood heating is incorporated in a central heating system, however, numerous other components apart from the heating boiler need to be selected and sized during the planning phase.

Figure 5.60 illustrates a complete heating system with a wood boiler. This system shows the most elaborate version possible, with various loads and a further heat supply provided by a solar thermal system.

The following section explains the functions of these system elements and provides notes on their sizing and execution.

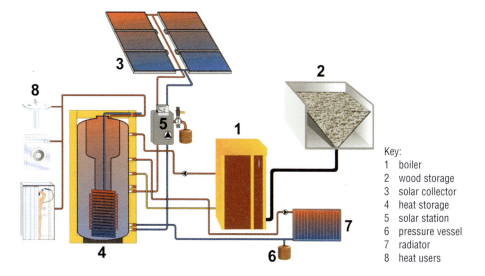

Figure 5.60.
Components of a combined pellet–solar thermal system
Graphic: www.wagner-solartechnik.de

Key:
1 boiler
2 wood storage
3 solar collector
4 heat storage
5 solar station
6 pressure vessel
7 radiator
8 heat users

Figure 5.61.
Pellet boiler connection pipes
Photo: www.gilles.at

5.4.1 Wood boilers

The choice and sizing of wood boilers have already been discussed in the previous sections. The descriptions and technical specifications in this section list all the information on outputs, types of installation and structural requirements.

As well as careful maintenance and proper servicing, an important criterion for trouble-free boiler operation is the protection of the boiler against corrosion. Corrosion appearing in the boiler's flue pipes is usually caused by condensing combustion products.

Each firing produces water in the form of steam as part of the oxidation process. If this comes into contact with cold surfaces, it condenses. If sulphur dioxide is also present in the flue gas, this can lead to the formation of aggressive sulphuric acid (H_2SO_4) in the condensate. This often causes considerable corrosion of the boiler walls or the flue system.

This corrosion problem can be avoided by using a suitable system design that maintains a high enough return temperature. A thermostat-controlled return flow of the hot water in the boiler maintains the temperature of the boiler wall at a level at which condensation cannot occur.

The tendency for condensation to form is determined by the water content of the fuel and the composition of the flue gases. The graph in Figure 5.62 shows the condensation points of the flue gases for various air/fuel ratios (l) and fuel water contents.

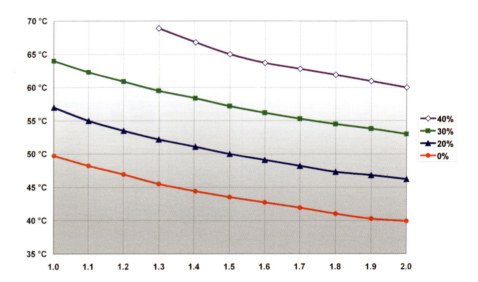

Figure 5.62.
Condensation points relative to the air/fuel ratio and humidity
Graphic: www.sesolutions.de

Flue gases from wood combustion with an air/fuel ratio $\lambda = 1.5$ and a humidity of around 20% condense on surfaces that are colder than 50°C. If the water content of the wood is around 40%, the condensation point with the same basic conditions occurs at 65°C.

In older houses the heating system is often designed with a flow temperature of 90°C and a return temperature of 70°C. Here most of the time there is no risk of condensation when heating with damp wood.

With modern radiators and boiler systems, however, low temperature levels are often chosen. For instance, 75°C for the flow temperature and 55°C for the return temperature of the heating system are usually chosen. This is completely within the condensation range of the flue gases, so that without thermostat-controlled increase of the return temperature, the boiler and flue system would quickly corrode.

Opening a thermostat vent ensures that the boiler quickly reaches the right temperature when firing up and that, during daily operation, it is not fed with return water that is too cold. It should always be ensured that a boiler monitoring system is provided to prevent corrosion damage when firing the boiler up and down.

5.4.2 Space heating demand

The building's space heating demand provides the basis for sizing the boiler output. This in turn provides the basis for selecting suitable boiler types. Because this is so important, the determination of this heat demand was explained in detail in section 5.2.

There are many different types of modern radiator (Figure 5.63). They have different radiation behaviours per square metre of radiator surface. If the heat demand of a room is determined using the procedure explained in section 5.2, this enables the optimum radiator size to be measured using the manufacturers' data.

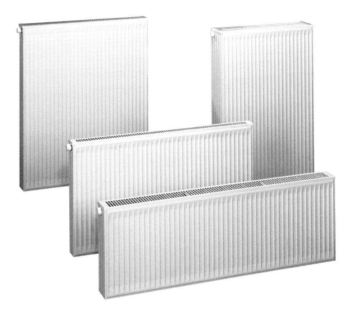

Figure 5.63.
Modern radiators
Photo: www.viessmann.de

5.4.3 Domestic hot water demand

The heat demand for heating domestic hot water in appliances such as showers, baths and washbasins can be taken to be an average of 12.5 kWh/m^2 of living area per year.

Table 5.24 provides key figures for the duration of use, the water withdrawal and the heat demand for sanitary appliances.

Table 5.24.
Domestic hot water demand
Data: Wamsler

Sanitary appliance	Length of use (min)	Extraction (l)	Appliance demand (kWh)
Bath tub (1600 mm × 700 mm)		140	5.8
Bath tub (1700 mm × 750 mm)		160	6.5
Bath tub (1800 mm × 750 mm)		200	8.7
Seated bath tub for small rooms		120	4.9
Shower cubicle with normal shower head	6	40	1.6
Shower cubicle with luxury shower head	6	75	3
Shower cubicle with luxury shower with 1 top and 2 side shower heads	6	100	4.1
Additional single shower heads	6	30	1.15
Wash basin	4	17	0.7
Bidet	10	20	0.8
Hand wash basin	4	9	0.35
Kitchen sink	10	30	1.15

This information provides the basis for the design of heating systems in buildings whose hot water use does not conform to the norm, such as guest houses and hotels.

5.4.4 Hot water storage tanks

The required content of the storage tank for wood boilers, where a solar thermal system is not used, is calculated in accordance with the European standard EN 303–5 using the following formula:

$$V_{St} = 15 T_C Q_N \left(1 - 0.3 \frac{Q_H}{Q_{min}} \right)$$

where V_{St} is the storage tank content (litres), T_C is the combustion time with a rated heat output (hours), Q_N is the nominal heat output (kW), Q_H is the determined heat load of the building (kW), and Q_{min} is the lowest heat output of the boiler (kW).

Note: In order to get a correct result, rooms such as guestrooms or hobby rooms that are not heated all the time must be deducted from the heat load of the building, Q_H.

Example: Calculating a storage tank
A large family home with a determined building heat load, Q_H, of 22 kW is to be equipped with a suitable storage tank.

The rated heat output, Q_N, of the boiler being installed is 26 kW, and its lowest heat output, Q_{min}, is 13 kW. When the boiler is filled with wood logs, the boiler manufacturer specifies a combustion time, T_C, of 4 h at the rated heat output.

$$V_{St} = 15 T_C Q_N \left(1 - 0.3 \frac{Q_H}{Q_{min}} \right)$$

$$= 15 \times 4\ h \times 26\ kW \times \left(1 - 0.3 \frac{22\ kW}{13\ kW} \right)$$

$$= 768\ \text{litres}$$

So the minimum storage tank size is 768 litres. On no account should the size fall below this. A good selection in this case would be a storage tank with a volume of 1000 litres.

5.4.5 Solar thermal systems

In most cases it is sensible to couple the central heating system with a solar thermal installation. This enables the building to be completely heated in an environmentally friendly manner with solar heat. This is achieved on the one hand through the collected solar radiation and on the other hand through the solar energy stored in the wood (Figure 5.64). A solar thermal system provides the building owner with added convenience and economic advantages as the solar radiation replaces the wood fuel.

A well-planned solar thermal system can save 50% of the annual wood demand. This is not only financially attractive; it also provides the operator with many advantages because there is no requirement to procure and prepare the wood fuel and dispose of the ash.

Solar thermal systems can generally be installed on any roof or facade that faces south. Figure 5.65 shows the variation in the yield of solar collectors when they deviate from a directly southern orientation.

Specific information on designing systems can be found in the DGS handbook on solar thermal systems. For the initial sizing of systems, the reference figures in Table 5.25 can be used.

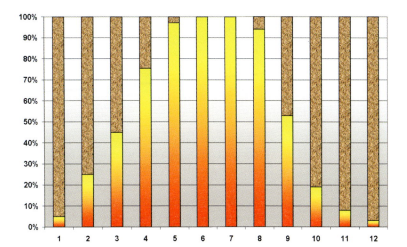

Figure 5.64.
Heat distribution with a pellet–solar thermal combination
Graphic: www.sesolutions.de

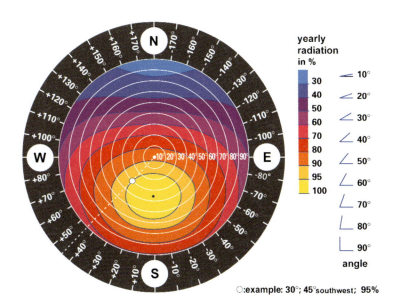

Figure 5.65.
Individual solar thermal yield
Graphic: www.viessmann.de

Figure 5.66.
Vacuum-tube collector
Graphic: www.viessmann.de

Table 5.25.
Average collector dimensions per household size

No. of residents	Collector area		Solar storage tank volume (litres)
	Flat-plate collector	Evacuated tube collector	
1–2	3.0–4.0	2.0–2.5	250–300
3–4	5.0–6.0	3.5–4.5	350–400
5–6	7.0–9.0	5.0–7.0	450–550
7–8	10.0–12.0	7.5–9.0	600–800

The storage tank volumes for solar thermal systems can be calculated using the demand on storage tanks for wood boilers. This enables investment costs to be saved, and achieves better efficiency of the overall system package.

5.4.6 Circulation pumps

Circulation pumps (Figure 5.67) are used to maintain the water flow in the heating system. Depending on the size and height difference of the heating system, the pumps must be able to cope with different delivery rates and water pressures. To size the pumps it is necessary to have a characteristic curve of the heating network. This can be determined mathematically using the following ratio:

$$\frac{H_1}{H_2} = \left(\frac{p_1}{p_2}\right)^2$$

If this system characteristic curve (A) is applied to the pump characteristic curve (B) provided by the manufacturer (Figure 5.68), the resulting delivery rate of the pump under these operating conditions can be derived from the intersection point (C) of the two curves.

For some applications it is necessary to increase either the displacement volume or the displacement pressure of the circulation pumps. This can be achieved easily by arranging the pumps in parallel or series.

The displacement volume of identical pumps can be increased if they are switched on in parallel. The new displacement flow, while retaining the same displacement pressure, is determined by adding the two individual displacement flows.

The displacement pressure can be increased with identical pumps when these are switched on in series. Here the new displacement pressure, while retaining the same displacement volume, is determined by adding the two displacement heights.

Figure 5.67.
Circulation pump for central heating systems
Photo: www.viessmann.de

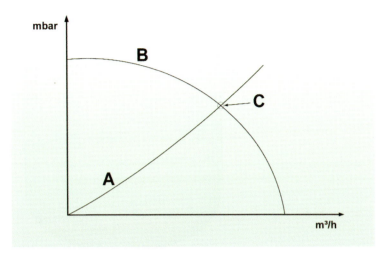

Figure 5.68.
Characteristic curve of a circulation pump
Graphic: www.sesolutions.de

5.4.7 Safety equipment for heating systems

The heating system should be equipped with, as a minimum, the following elements:

- closed expansion tank
- safety vent at the highest point of the boiler
- safety plug
- thermometer
- manometer
- self-acting device for heat removal, activated if the operating temperature is exceeded.

Figure 5.69.
Security installations
Photo: www.viessmann.de

5.4.8 Expansion tanks

Water expands when it is heated. This natural phenomenon would cause considerable pressure fluctuations in a closed heating system if the temperature of the system changed.

Therefore, in order to avoid damage to the pipe network, boiler and radiators, an expansion tank is installed in every closed water system (Figure 5.70). This consists of a steel-cased rubber bubble that is filled with an inert gas such as nitrogen and can therefore absorb pressure fluctuations in the network.

Figure 5.70.
Expansion tank for central
heating systems
Photo: www.viessmann.de

The size of the expansion tank is crucial for the safety and pressure resistance of the entire pipe network. Expansion tanks are designed using the following formula:

$$V_T = f\left[\,(V_S + V_B)\,E_f + 2.4\,\right]$$

where V_T is the total expansion tank volume, f is an expansion factor (= 2 for expansion tanks), V_S is the system volume (including the hot water storage tank), V_B is the volume of the boiler water, and E_f is an expansion factor for hot water.

Note: When determining the system volume, V_S, the volume of the existing storage tank in the heating system should be taken into consideration.

The expansion factor, f, for the cold heating water used when filling the system can be determined from the graph in Figure 5.71. A maximum water temperature in the heating circuit of 82°C produces an expansion factor, f, of 2.9%.

The next important parameter for sizing the expansion tank is the volume of water in the heating system, V_S. It can be determined for the most common heating systems from the graph in Figure 5.72.

With special buildings it is also possible to estimate the heating volume, V_S, using the existing pipe and radiator dimensions (see Tables 5.26–5.30).

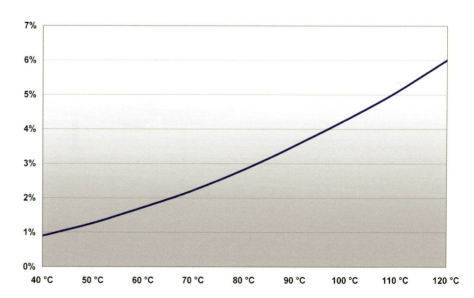

Figure 5.71.
Expansion curve of water
Graphic: www.sesolutions.de

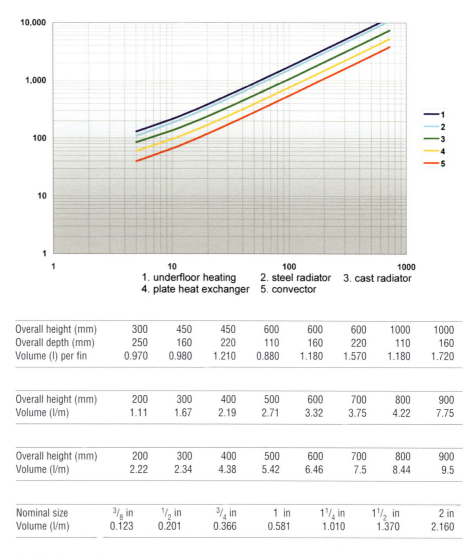

Figure 5.72.
Water content of typical central heating systems
Graphic: www.sesolutions.de

1. underfloor heating 2. steel radiator 3. cast radiator
4. plate heat exchanger 5. convector

Table 5.26.
Steel hospital-type radiators

Overall height (mm)	300	450	450	600	600	600	1000	1000
Overall depth (mm)	250	160	220	110	160	220	110	160
Volume (l) per fin	0.970	0.980	1.210	0.880	1.180	1.570	1.180	1.720

Table 5.27.
Steel single panel radiators

Overall height (mm)	200	300	400	500	600	700	800	900
Volume (l/m)	1.11	1.67	2.19	2.71	3.32	3.75	4.22	7.75

Table 5.28.
Steel double panel radiators

Overall height (mm)	200	300	400	500	600	700	800	900
Volume (l/m)	2.22	2.34	4.38	5.42	6.46	7.5	8.44	9.5

Table 5.29.
Threaded pipes

Nominal size	$3/8$ in	$1/2$ in	$3/4$ in	1 in	$1 1/4$ in	$1 1/2$ in	2 in
Volume (l/m)	0.123	0.201	0.366	0.581	1.010	1.370	2.160

Table 5.30.
Copper pipes

Nominal size (mm)	8 × 0.8	10 × 0.3	12 × 1	15 × 1	18 × 1	22 × 1.2	28 × 1.2
Volume (l/m)	0.030	0.060	0.080	0.130	0.200	0.300	0.520

Example
A building has a heating system volume, V_S, of 725 litres and a boiler volume, V_B, of 25 litres. The maximum temperature of the heating system was measured at 82°C. From this a factor E_f of 2.9% was determined (Figure 5.71).

How large should the expansion tank be?

$$V_T = f \left[(V_S + V_B) E_f + 2.4 \right]$$

$$= 2 \times \left[(700 \times 50) \times 2.9 + 2.4 \right]$$

$$= 48.30$$

The calculated volume of the expansion tank is 48.30 litres. An expansion tank should therefore be selected that has an overall volume of 50 litres.

5.4.9 Soundproofing

Biomass boilers that are not correctly designed can cause noise pollution in buildings. Here it is primarily the fans, augers and pneumatic conveyor systems that produce noise.

If the following points are taken into consideration at the planning stage, it is possible to avoid noise problems with biomass heating:

- Adapting the architecture of the building. Bedrooms should not be arranged directly above the boiler room. Sound-transmitting chimneys should not pass by bedrooms.
- Acoustic separation within new buildings. Boiler rooms and storerooms in new buildings should be separated from the walls and concrete floors by using elastic absorption material.
- Vibration separation of mechanical parts. All contact points between mechanical parts and wall and floor materials should be acoustically insulated or isolated with vibration dampers.
- Installation of sound reduction elements recommended by the manufacturer. At the final inspection it should be ensured that all sound reduction elements specified by the manufacturer, such as isolation mats, have been properly installed.
- Sound emissions can be compared to other reference sites. The lack of binding standards means that it is often only possible to control and assess the sound emissions by visiting reference systems.

It is not just the operation but also the delivery and unloading of fuel that causes sound emissions. A suitable storage location should be chosen that, coupled with acceptable delivery times, ensures that no conflicts arise with other residents.

5.5 Chimneys

It is essential that chimneys are correctly sized and designed, as this determines the output when sizing small combustion systems. This is because the natural draught of the chimney removes the flue gases from the boiler while at the same time sucking fresh air into the boiler, which is essential for the combustion process.

The chimney effect is based purely on the physical lifting capability of hot air. Figure 5.73 shows the pressure ratios of boiler systems with and without induced-draught fans. Systems with positive pressure combustion (a) such as oil and gas boilers, which have forced-air burners, are capable of overcoming the pressure loss of the boiler interior through their fans.

Boilers with under pressure combustion (b), such as most wood-fired boilers, rely completely on the natural draught of the chimney to discharge the flue gases and to supply the boiler with fresh air. In many cases this is supported by an induced-draught fan that can improve the draught output of the chimney by creating an artificial draught. This also enables the combustion output of a negative pressure boiler to be varied.

A correctly sized chimney flue height is fundamental for increasing the output of boilers; the chimney flue must be tailored as precisely as possible to the boiler. Too strong a flue draught leads to increased standby losses, thus reducing the efficiency of wood-fired boilers.

To improve the balance between the boiler and the existing chimney, it is generally sensible and economic to use a draught limiter (draught regulator). It is certainly necessary to install a draught limiter with a flue draught of 20 mbar or more.

Modern draught limiters are usually equipped with a pressure-relieving flap that is ripped open if there is a blowout in the chimney. This enables the resulting pressure to be released, protecting the house and chimney from damage.

Chimney systems for small combustion systems with wood must fulfil several criteria:

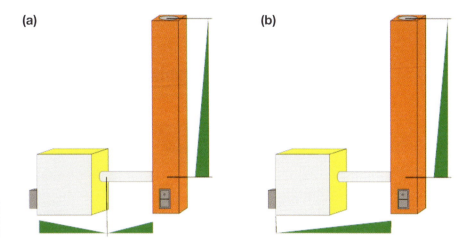

*Figure 5.73.
Boiler with positive and negative
pressure combustion
Graphic: www.paradigma.de*

- Temperature resistance. The chimney system must resist temperatures up to 400°C.
- Soot fire resistance. Chimney systems must have a soot fire resistance of up to 1200°C.
- Moisture resistance. The chimney systems must be completely moisture resistant, because flue temperatures below 160°C can occur.
- Surface finish. The chimneys must have a smooth inner surface without cracks.
- Thermal insulation. The chimneys must have suitable thermal insulation to prevent flue gases from condensing on cold chimney surfaces.
- Cross-section. The chimney must have a constant-sized cross-section that does not narrow: that is, the diameter must not change.

Two different chimney systems that fulfil these criteria are mainly used for wood combustion: ceramic chimneys and stainless steel chimneys. See Figure 5.74:

*Figure 5.74.
Chimney systems made of ceramic
material and stainless steel
Photos: www.pro-schornstein.de;
www.viessmann.de*

Chimney designs are based on the following main criteria:

- height above sea level of the place of erection
- type of planned hearth appliance
- multiple use of the chimney
- cross-sectional form
- diameter
- effective chimney height
- length in cold area
- length outside.

Figure 5.75 shows the different lengths.

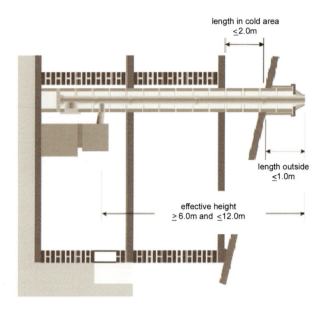

*Figure 5.75.
Chimney dimensions
Graphic: www.schiedel.de*

All boiler manufacturers specify the draught requirements for the chimney in their technical installation instructions. These vary depending on the technical design of the boiler.

By way of example, Figure 5.76 shows the permitted chimney shapes and draught heights for a Paradigma boiler according to German provisions for chimneys used for wood-fired heating.

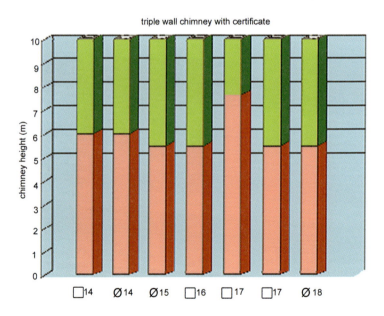

*Figure 5.76.
Provisions for chimneys
Graphic: www.paradigma.de*

5.5.1 Chimney flue pipes

The flue pipes connected to the chimney should be short (length < 2.0 m) and airtight. The connection to the chimney must always be laid so that it slopes upwards (>15°). In practice, upward tilts of 30–45° in the direction of flow have proven to be reliable. Furthermore, the connection should be insulated (> 5 cm, heat resistant) and, if possible, executed without any bends. The entry to the chimney should facilitate the flow and curve upwards. See Figure 5.77.

A certain amount of positive pressure is to be expected when wood boilers are fired up. For this reason, the flue pipe to the chimney should be laid so that it is completely airtight. Here it is possible to use temperature-resistant silicone as sealing compound; alternatively, the flue pipe can be welded tight.

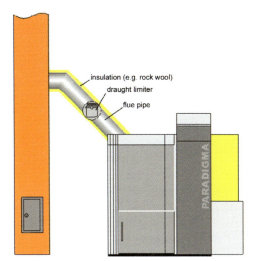

Figure 5.77.
Chimney connection of a pellet boiler
Graphic: www.paradigma.de

Furthermore, it is sensible to lay a flexible and insulated flue pipe to the chimney to improve the noise insulation. The flue pipe should never be bricked into the chimney, as this could otherwise cause noise insulation problems. The flue pipe should always have an airtight inspection opening to enable the soot deposits to be controlled and, if need be, removed.

5.6 Storage

5.6.1 Stores for wood logs

The most important criterion for wood logs is the degree of dryness. It has already been mentioned in section 5.2 that damp wood can lead to corrosion damage in wood boilers. In addition, each litre of water that is removed from the wood as steam uses 0.7 kWh, which is discharged unused through the chimney and is not available for space heating.

Other problems to do with damp wood result from the temperature dropping during combustion, so that the fire zone does not produce the necessary heat. This leads to the risk of unburned wood gas collecting as creosote and soot in the flue valves and the chimney.

All these problems can be avoided if the wood is dried properly; a two-year drying period outside is ideal. After this drying, almost all chopped wood logs have a water content that is suitable for combustion. See Figure 5.78.

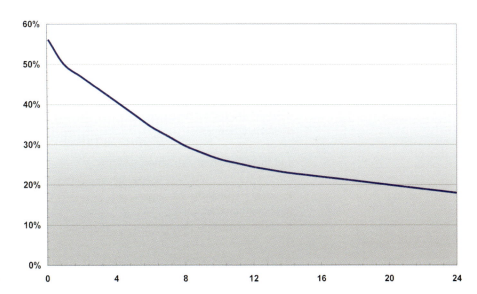

Figure 5.78.
Dry curve of wood
Graphic: www.sesolutions.de

In order to attain this condition the wood needs to be stored correctly. The following basic conditions should be observed:

- The wood should be sawn and split so that it is ready for use and stored.
- The wood should lie on a base roughly 20 cm high that is permeable to air.
- There should be a vertical air gap 5–10 cm wide behind the wood stack.
- The wood stack should be protected against rain with a roof cover.
- Do not completely cover the wood with plastic sheeting.
- Store wood only in rooms with a sufficient air supply, as otherwise there is a risk of mould forming.

Note: Wood logs have a relatively low energy density: 5 m^3 of wood logs are required to replace 1 m^3 of fuel oil. If it is intended to heat throughout the year with wood logs, it is important that you always store at least 1.5 times the annual fuel demand in order to enable the wood to dry properly and to cover cold periods. Such an amount of wood corresponds to a storage volume of at least 7.5 m^3.

The storage regulations in some countries or regions may stipulate that the wood should not be stored outside but only in special wood stores. Figure 5.79 shows such a wood store.

Figure 5.79.
Storing wood logs
Photo: www.kuenzel.de

5.6.2 Possibilities for storing pellets

There are essentially six different possibilities for storing pellets, which are tailor-made for different spaces. In technical terms these consist of four different storage units:

- sack silos
- pellet bunkers
- underground storage tanks
- storage hoppers

and three different extraction systems:

- auger extraction
- vacuum extraction
- static extraction (hopper).

All the systems described here can ensure trouble-free extraction of pellets from the respective storage device. The various combinations of these processes are, however, developed for different application areas.

5.6.2.1 BASIC CONSIDERATIONS FOR PELLET STOREROOMS

The pellets are delivered by silo trucks and are injected into the storeroom. The silo trucks generally have a pump hose with a maximum length of 30 m. Therefore the pellet storeroom (or the fill coupler for the sack silo) should not be more than 30 m from the house entrance.

Even if it is to be expected that the pump hose lengths will be longer, you should still consult your preferred pellet supplier to sort out the technical possibilities before adopting any solutions that require greater hose length. You should also bear in mind with special solutions that it may not be possible to change suppliers should you wish to do so for price or quality reasons.

5.6.2.1.1 LOCATION OF THE BOILER ROOM

If possible, the pellet storage room should adjoin an external wall, as the fill pipe must be accessible from outside. With internally situated storage rooms, the feed and flue pipes must be run to the external wall. The pellet boiler must always be switched off before filling the pellet store.

If possible, the boiler room should also adjoin the external wall, to ensure a direct supply of combustion air for the pellet boiler. With internally situated boiler rooms, a supply duct must be run from the boiler room to the external wall.

5.6.2.1.2 SIZING THE PELLET STOREROOM

The pellet store should always be rectangular, and the room should not be wider than 2.0 m if possible. Dimensions such as 2 m × 3 m or 1.8 m × 3.2 m have proved reliable in practice. The flow mechanics of the wood pellets mean that the narrower the room, the smaller the size of the non-usable voids.

The size of the storeroom depends on the heat demand of the building. It should be large enough, however, to store a year's supply of fuel. The following rules of thumb apply when sizing pellet storerooms:

- 1 kW heat load = 0.9 m^3 store room (including void).
- Usable storage space = 2/3 storage room (including void).
- 1 m^3 pellets = 650 kg.
- Energy content = 5 kWh/ kg.

Example
Single-family home with a heat load of 15 kW = annual demand of pellets of 5800 kg

15 kW heat load × 0.9 m^3/kW = 13.5 m^3 store room volume

Usable room volume = 13.5 m^3 × 2/3 = 9 m^3

Storable amount of pellets = 9 m^3 × 650 kg = 5850 kg

Required storage room volume = 13.5 m^3 ÷ 2.4 m (room height) = 5.6 m^2 area

A good storage room area would be 2m × 3m = 6m^2. This should not be exceeded.

Stored energy amount = 5850 kg × 5 kWh/ kg = 29,250 kWh

5.6.2.1.3 DAMP PROTECTION

Pellets are highly hydroscopic: that is, they absorb water. If they come into contact with damp floors or walls they expand and break up, and are therefore unusable.

The following technical requirements therefore apply for pellet stores:

- The pellet store must remain dry throughout the year.
- Normal air humidity, such as that which occurs throughout the year in normal dwellings as a result of the weather, does not harm the wood pellets.

- Should there be a risk from occasionally damp walls (e.g. in old buildings) the walls should be clad with ventilated timber panelling. Alternatively, there is the possibility of storing the pellets in industrial pellet stores such as grain silos.
- New buildings in which humidity is still trapped in the walls following construction should be completely dried out before installing a pellet store.

5.6.2.2 STORING AND TRANSPORTING WOOD PELLETS

The flowchart in Figure 5.80 enables the most suitable pellet storage system for a building to be chosen quickly.

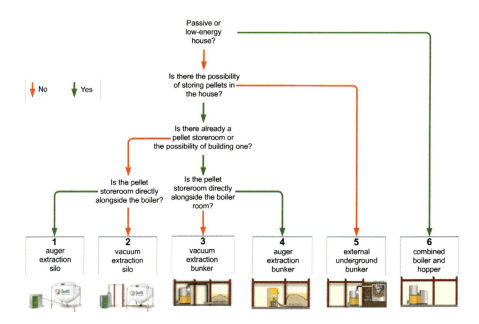

Figure 5.80.
Choice of pellet stores
Graphic: www.sesolutions.de

The following pages describe the method of functioning and technical features of these storage types.

5.6.2.3 SYSTEM 1: AUGER EXTRACTION FROM A PELLET SILO

System 1, which uses a pellet silo with an auger, is suitable when it is possible to install the pellet boiler and the store directly alongside each other in an internal room of the building.

The main part of the system is the pellet silo. This consists of a wear-resistant, artificial fabric that is hung like a tent within a stable, steel frame construction (Figure 5.82).

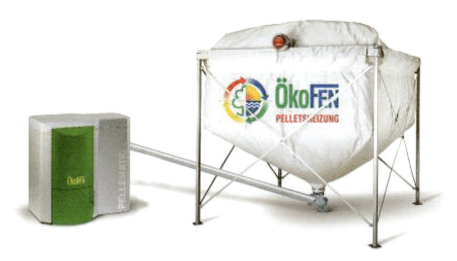

Figure 5.81.
Pellet storage in a sack silo
Photo: www.oekofen.at

Figure 5.82.
Sack silo in a tubular frame
Photo: www.paradigma.de

On the front of the completely closed silo is a coupler pipe that is connected to the support frame. The pellet hose from the tanker is connected to this. This enables the silo to be filled by a conventional pellet lorry.

The fabric material of the pellet silo has the characteristic of being permeable to air but dustproof. This means that these silos do not require a separate exhaust pipe. During the filling process described on the following pages, the silo inflates, the injected air escapes through the fabric material. and the pellets are left behind in the silo (Figure 5.83).

Figure 5.83.
Filling a sack silo
Photo: www.paradigma.de

Once the pellet silo has filled, you first have to wait for the accumulated air to escape. The pellet boiler can then be operated again as normal. The pellets are removed from the floor of the silo through a round opening and conveyed with the flange-mounted auger to the boiler.

Auger conveyors for wood pellets basically come in two different designs: straight or bent. Straight pellet conveyors consist of a pipe onto which the screw blade is welded. With bent pellet augers, both parts are connected together with a universal joint.

Pellet silos with auger extraction are reliable, low-maintenance systems for the uncomplicated storage of wood pellets in boiler rooms.

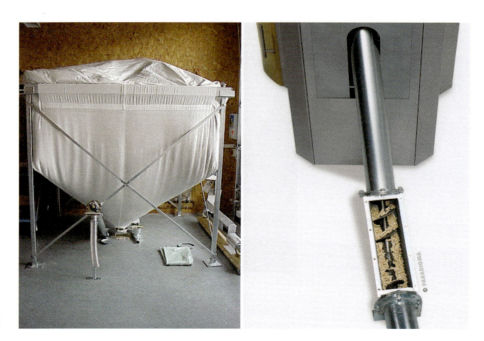

Figure 5.84.
Auger extraction from the silo
Photo: www.paradigma.de

5.6.2.4 SYSTEM 2: VACUUM EXTRACTION FROM A PELLET SILO

System 2 with a pellet silo and vacuum extraction (Figure 5.85) is suitable when it is not possible to install a permanent store inside the building, and the pellet boiler and storage hopper must be in separate spaces.

Pellet silos using vacuum extraction are technically identical to silos using auger extraction. To install the vacuum extraction system, a suction piece is attached to the exit hole at the bottom of the silo in place of the auger flange.

Two spiral hoses are fixed to this special socket piece. The pellets are drawn up using a suction motor through one of the spiral hoses into the pellet boiler. The sucked-in air is then returned to the silo via the other hose, where it can again be used for transferring the pellets.

Vacuum systems can cover distances of up to 20 m between the pellet store and the boiler. Here it should be taken into consideration when running the conveyor hoses through ceilings that two approximately 60 mm-wide spiral hoses must be laid (flow and return hoses). Even if it possible to lay the hoses separately, the hose lengths should not differ by more than 10%.

When laying the hoses, the bend radii of the hoses must not be not smaller than five times the external diameter of the hoses. This means that 60 mm spiral hoses require a bending radius of 5×50 mm = 300 mm.

If they have to overcome gradients, lengths no greater than 3 m should be laid at an incline. If horizontal hose sections are installed that are at least 1 m long, it is also possible to combine sloping sections.

Figure 5.85.
Sack silo with vacuum extraction
Photo: www.oekofen.at

When installing and operating the system it is important to ensure that the hoses and connections are absolutely airtight. Here hose clamps specified by the manufacturer should be used, which provide a potential equalization connection for the copper wire embedded in the elastic hoses.

5.6.2.5 SYSTEM 3: VACUUM EXTRACTION FROM A PELLET BUNKER

System 3 is identical in terms of the vacuum technology used in system 2 described above. Instead of using a prefabricated pellet silo, the wood pellets are stored in their own specially equipped store. See Figure 5.86.

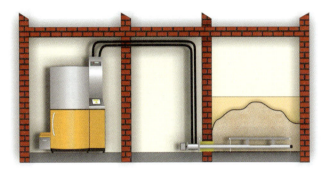

Figure 5.86.
Pellet bunker with vacuum extraction
Graphic: www.wagner-solartechnik.de

The system is particularly suitable for homeowners who wish to provide a pellet storage room in their buildings but cannot do this in the immediate vicinity of the pellet boiler room. When laying suction pipes in the house it should be taken into consideration that the tubes can give off clattering noises during the transfer of the pellets, which takes place roughly twice a day.

There are numerous ways in which the vacuum system can be connected to the built-in pellet storage room. It can either be coupled with a centrally installed auger or connected to three or four suction hoses whose nozzles are laid on the floor of the pellet store. (Figure 5.87). Here the pellet auger empties the store in the middle and conveys the wood pellets in the air current created by the suction motor.

Figure 5.87.
Auger extraction with suction hoses
Graphic: www.hargassner.at,
Photo: www.paradigma.de

The other possibility is to install suction hoses on the floor of the pellet storeroom (Figure 5.88). Here the circulating air draws the pellets into the nozzles of the suction hoses where they are sucked with the flow current to the boiler. These suction hoses are screwed to the floor of the storerooms at intervals of 50–75 cm. For this purpose the nozzles of the suction hoses are equipped with floor panels with drill holes.

Figure 5.89 shows a pellet bunker in which the pellets are removed with suction hoses. Characteristically shaped pits form directly above the hoses. Although pellets are a bulk material they do not flow like water from a tank. Instead they slide

Figure 5.88.
Vacuum inlets at the bottom of a pellet bunker
Photo: www.windhager-ag.at

downwards and, if the pellets are removed correctly, they form an inverted cone. In this way pellet bunkers with sloping floors can be completely emptied.

Figure 5.89.
Vacuum inlet and pellet bunker in action
Photo: www.windhager-ag.at;
www.paradigma.de

5.6.2.6 SYSTEM 4: AUGER EXTRACTION FROM A PELLET BUNKER

System 4 is identical in terms of the structure of the pellet store used in system 3 described above. However, instead of transporting the wood pellets with a vacuum system, the pellets are extracted from the storeroom with an auger before being transported the rest of the distance to the pellet boiler with another screw conveyor (Figure 5.90). Thus the mechanism used by this system for conveying pellets to the pellet boiler is identical in technical terms to system 1.

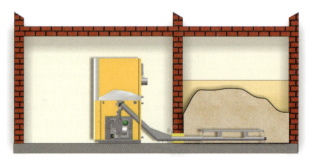

Figure 5.90.
Pellet bunker with auger extraction
Graphic: www.wagner-solartechnik.de

If the pellet storeroom is to be built within a solid structure, a closed room should be equipped with a support structure that converts the storage room into a pellet store. Here sloping floors on supporting frames must be installed in accordance with the cross-section shown in Figure 5.91.

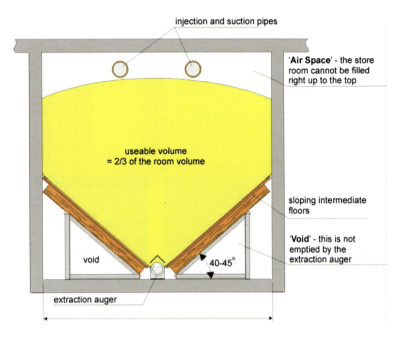

Figure 5.91.
Cut through of a pellet bunker
Graphic: www.depv.de

Figure 5.92 shows the floor plan of an ideal pellet storeroom. The longitudinal section through the pellet storeroom is designed as shown in Figure 5.93.

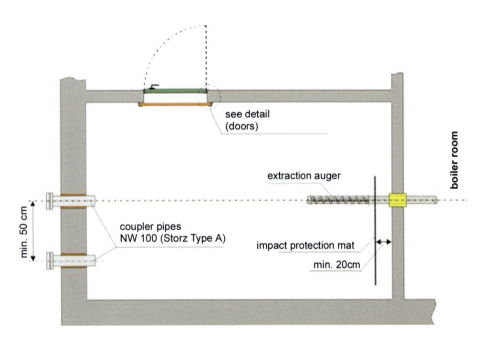

Figure 5.92.
Cross-section of a pellet bunker
Graphic: www.depv.de

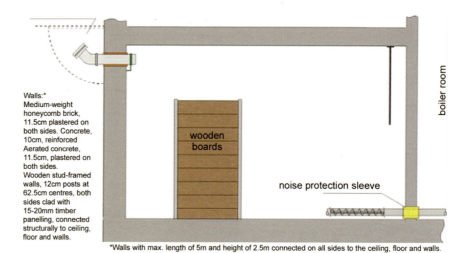

Figure 5.93.
Longitudinal section of a pellet bunker
Graphic: www.depv.de

5.6.2.6.1 Structural requirements for the pellet store

The surrounding walls must be able to withstand the loads imposed by the pellets (bulk density 650 kg/m^3). The following wall depths have proven themselves in practice:

- medium-weight honeycomb brick, 11.5 cm plastered on both sides
- concrete, 10 cm, reinforced
- aerated concrete, 11.5 cm plastered on both sides
- brick, 12 cm, plastered on both sides
- wooden stud-framed walls comprising 12 cm posts at 62.5 cm centres, both sides clad with 15–20 mm timber panelling, connected structurally to ceilings, floor and walls.

These requirements apply to walls with a maximum length of 5 m and a height of 2.5 m, which are connected on all sides to the ceiling, floor and walls.

There are generally no fire protection requirements for pellet store doors or hatches for stored amounts up to 15,000 kg. Door and hatches must open outwards and have a continuous seal (dustproof).

The inside of doors and hatches to pellet stores must be protected with wooden boards to prevent pellets from pressing against the doors or hatches. The door handle should be removed from the inside. The door lock should be sealed on the inside so that it is dustproof. This can be done with strong insulation tape.

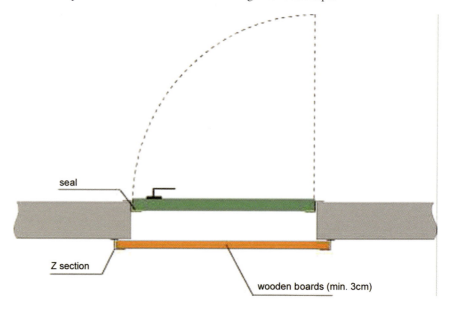

Figure 5.94.
Detail of a pellet bunker door
Graphic: www.depv.de

The impact protection mat (e.g. 1250 mm × 1500 mm) is designed to protect the pellets from being destroyed when they impact against the surrounding walls. It also protects the walls themselves from being damaged. The impact protection mat consists of abrasion- and age-resistant rubber material with mounting brackets to fasten it to the ceiling. It is hung opposite the fill tubes with a 20 cm gap between it and the rear wall.

5.6.2.6.2 Filling system

In order to fill the pellet store two fill tubes with couplings are required; a suction fan is connected to one of the fill tubes when fuel is delivered.

To install the fill tubes it is necessary to create holes in the walls with 125–1500 mm diameters. Here, polyethylene or PVC pipes built into the wall have proven reliable in practice. The fill coupling and pipes should be firmly fixed so that the fill pipes do not twist round when the pellet supplier attaches the hose to the coupling. When assembling the heating system, waterproof polyurethane foam is injected into the gaps between the fill pipes and the prepared wall openings.

Note: The fill pipes must be earthed (1.5 mm^2 wire connection to the equipotential bonding terminal). This earthing is necessary to prevent electrostatic charges occurring during the filling process.

Straight and curved pipes should be beaded at both ends to enable them to be connected together with expansion rings so that they are firm and dust-tight.

When installing the filling system, you must not use:

- pipes made of plastic (danger of electrostatic charges)
- pipes whose structure could destroy the pellets during the filling process (e.g. folded spiral-seam tubes used in ventilation systems).

General notes:

- Only metal pipes may be used for the filling system.
- The filling system must be earthed against electrostatic charges.
- The filling systems used must be continuously smooth on the inside. Suitable are, for example, socketless cast iron pipes (SML pipes), as these can be joined together with flush connections using an appropriate clamp.
- If the pipe connections are welded, no edges or welds should protrude on the inside that could destroy the pellets.
- Bends should be avoided if at all possible. If bends have to be used, they should have as large a radius as possible. Alternatively, 90° changes in direction can also be achieved using two 45° bends.
- The filling system must not end with a bend. Instead the bend must be following by a straight piece that is at least 50 cm long to allow the material flow to settle.

5.6.2.6.3 Sloping floors

The storage room must have sloping floors that enable it to be almost completely emptied by the extraction system used.

General notes:

- The sloping floors must be at a 40–45° angle so that the pellets can flow down by themselves.
- The sloping floors should preferably be installed with timber panels with a surface that is as smooth as possible, such as chipboard or coated chipboard. OSB boards have a surface that is too rough.
- The sloping floors must be able to withstand the loads imposed by the pellets (bulk density 650 kg/m^3). A suitable substructure, for example, is shown in Figure 5.95.
- Suitable steel angles are available for the substructure that substantially ease the construction of the sloping floors.
- The sloping floors should extend right up to the surrounding walls to prevent pellets from trickling into the voids beneath (from where it is not normally possible to retrieve them).

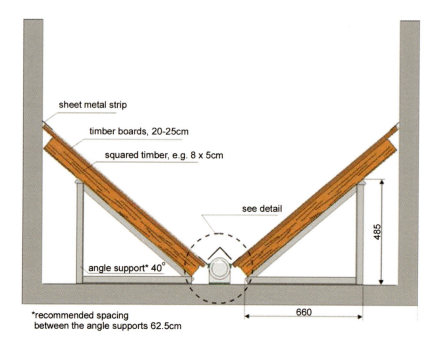

Figure 5.95.
Construction of the tilting floor
Graphic: www.depv.de

- The sloping floors should not reduce the lateral openings between the conveyor duct and the cover.

The sloping floors are terminated at the conveyor duct with sheet metal and sponge rubber strips.

Note: The sheet metal strips must abut the conveyor duct around 10 mm below its opening. This serves to take the pressure off the conveyor spiral caused by pellets collecting on the sides.

The junction of the sloping floors can vary in detail from the design illustrated below (see Figure 5.96). The manufacturer's specific assembly instructions must be observed. This detail should always be executed in accordance with the specifications of the respective firm manufacturing or supplying the conveyor system.

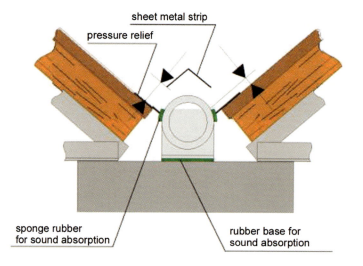

Figure 5.96.
Detail of the tilting floor
Graphic: www.depv.de

5.6.2.6.4 INSTALLATIONS IN THE PELLET STORE

Existing pipe connections and discharge pipes etc. that cannot be removed easily, and which could cross the path of the pellets during filling, should be clad to protect them from fracturing (for example by using metal deflection covers). It should be ensured that the deflection cover cannot damage the pellets (no rectangular cladding).

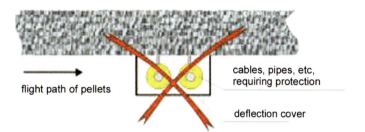

Figure 5.97.
Construction of inserts
Graphic: www.depv.de

5.6.2.6.5 ELECTRICAL INSTALLATIONS IN THE PELLET STORE

No electrical installations may be situated in the pellet store: this includes switches, lights and conduit boxes (exceptions are explosion-protected designs).

The house fuse box should be attached close to the fill couplings. It should contain a plug socket (230 V) for the pellet supplier's suction fan. In addition, it should be equipped with a door switch. This is primarily used to switch off the current supply to the boiler.

If the door of the fuse box is opened, the heating system switches itself off automatically and prevents any possible return suction of flue gases.

5.6.2.6.6 FIRE PROTECTION

The general national or regional fire protection provisions provide the basis for the fire protection requirements. If in doubt, you should contact the responsible local authorities.

Table 5.31.
Basis for fire protection requirements

Pellet storage amounts £15,000 kg 23 m^3	Pellet storage amounts ≥ 15,000 kg
No requirements for walls, ceilings, doors or their use	Walls and ceilings F90 No cables or ducts through walls No other use Door self-closing and fire-retardant (T 30)
Nominal heat output of boiler £50 kW (room in which heating appliance is installed)	Nominal heat output of boiler ≥ 50 kW (boiler room)
No requirements for the room Combustion air supply for the heating appliance min. 150 cm^2	Walls and ceilings F90 Doors self-closing, opening to the outside and fire-retardant (T 30)
Distance of heating appliance to fuel store 1 m or use radiation shield	No other use Ventilation ducts min. 150 cm2 (above 50 kW + 2 cm^2/kW)
Pellets up to 15,000 kg may be stored in the same room as the heating appliance	Up to 15,000 kg pellets may be stored Distance of heating appliance to fuel store 1 m or use radiation shield

5.6.2.6.7 PELLET STORE CHECKLIST

The checklist in Table 5.32 is intended to help installers check the basic requirements for individual pellet stores before filling them for the first time.

Table 5.32.
Pellet store checklist
Source: The German Wood Pellet Association DEPV

Electrical installation: There should be no lamps, switches, plugs, junction boxes, etc. in the pellet store (risk of explosion during injection).

Filling system: The fill couplers should be firmly fixed and attached so that they are easily accessible for the pellet supplier.
The coupler cover should be clearly labelled with the following notice:
'CAUTION! SWITCH OFF heating before filling the pellet store.'
The filling system must be earthed.

Impact protection mat: The impact protection mat has been hung on the opposite side of the store from the fill pipes.

Storeroom door: Existing doors or hatches in the storeroom have a continuous seal around them.
The keyhole has been sealed from the inside.
Protection boards have been installed on the inside of the door.

Floor and walls: The floor and walls of the storeroom are dry.

5.6.2.7 SYSTEM 5: EXTERNAL UNDERGROUND BUNKER

System 5 with underground bunkers (Figure 5.98) is used when owners do not have enough space for a pellet store in the building or wish to use existing rooms for other purposes.

Figure 5.98.
Underground bunker with vacuum extraction
Graphic: www.wagner-solartechnik.de

A storage tank made of concrete or plastic is buried in the earth at a distance of around 1 m from the wall of the house. Figure 5.99 shows the four main phases for installing an underground store for wood pellets.

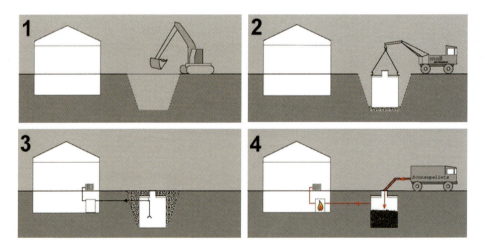

Figure 5.99.
Installing an underground bunker
Graphic: www.mallbeton.de

In phase 1 a pit must be excavated with a digger. As this store can have a maximum construction height of 3.00 m, the health and safety provisions for excavation work must be observed. This may mean that the pit will need to be secured with formwork to prevent it from collapsing.

On reaching the required depth, a layer of sand or lean concrete 10 cm deep must be laid in the pit to allow the pellet bunker to rest evenly so that it does not lie on any loose stones or boulders.

*Figure 5.100.
Installing an underground bunker
Photo: www.mallbeton.de*

When this excavation work is completed, in phase 2 the pellet bunker can be lifted into the pit with the crane from the lorry that delivered the bunker. The pit is then filled in around the pellet storage tank and the tank is covered with topsoil up to the edges of the feed duct. When filling in the pit in layers the earth must be sufficiently compacted by using a plate vibrator or rammer (frog-type jumping rammer). Otherwise the soil around the storage tank can settle unevenly. Figure 5.100 shows the storage tank being lifted into place

In phase 3 of the construction work the pellet store is equipped with the required internal fittings. As seasonally determined temperature fluctuations in the pellet store can cause condensation to form, the pellet store is equipped with a special locking system that can draw off condensation on the tank lid to the outside. Once all the equipment has been installed, the pellet boiler can be filled as shown in phase 4.

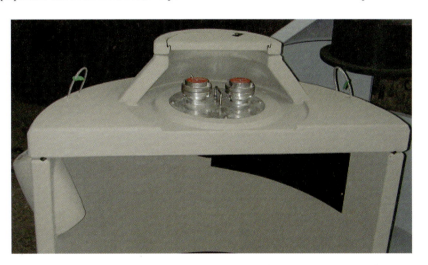

*Figure 5.101.
Cross-section through an
underground bunker
Photo: www.mallbeton.de*

Note: As with all pellet storage systems, the boiler must be switched off before filling; otherwise flue gas and burning heat could be sucked into the bunker.

To fill the underground storage tank, the supply and exhaust hoses are connected to the two pipes. The rest of the filling process is identical to that for the other pellet stores (Figure 5.102).

Pellet storage tanks made of concrete are permanently buried in damp soil. Therefore only containers made of water-resistant concrete may be used. Normal concrete products such as conventional tanks or cisterns often do not fulfil this criterion. Therefore only storage tanks that have been designed as pellet stores should be used; otherwise, if the pellets become too damp, this could damage the transport system, the pellet boiler, the flue pipes and the chimney.

*Figure 5.102.
Filling an underground bunker
Photo: www.mallbeton.de*

*Figure 5.103.
Water-resistant concrete
Photo: www.mallbeton.de*

5.6.2.8 SYSTEM 6: COMBINED BOILER AND HOPPER UNIT

System 6 with a combined boiler and hopper (Figure 5.104) is used primarily when buildings have particularly low heating demands. This can be the case, for example, with low-energy or passive homes that use less than 30 kWh/m^2 of living space per year.

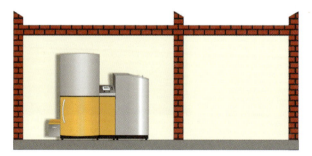

*Figure 5.104.
Combined boiler and hopper
Graphic: www.wagner-solartechnik.de*

The storage hopper adjoining the pellet boiler is loaded with sacks of wood pellets. Here, the cover of the storage hopper is opened and the pellets are poured in. The technical construction of the storage hopper is similar to that of a small pellet bunker with an auger and a pressure-relieving metal plate.

Because it is combined with a storage hopper, the pellet boiler has wider external dimensions, which should be taken into consideration when designing the internal spaces.

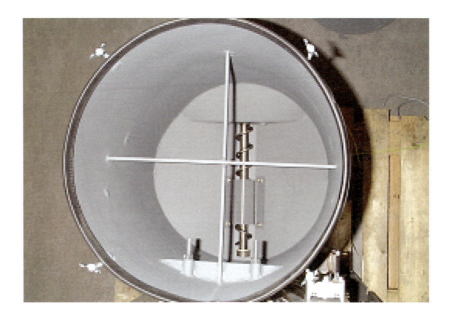

*Figure 5.105.
Auger extraction in the hopper
Photo: www.paradigma.de*

*Figure 5.106.
Pellet boiler with hopper
Photo: www.oekofen.de*

5.6.3 Storage possibilities for woodchips

Woodchips are much less homogeneous than wood pellets. For this reason they are generally moved using front loaders, tractors with front shovels or other digger-like machines. These procedures demand quite different structural requirements for a storage bunker than is the case with wood pellets.

In contrast to pellet bunkers, which can be loaded via hosepipes from distances up to 30 m, with woodchip bunkers it must be possible to drive directly up to them with the aforementioned machines. This technical requirement means that there are six possible variants for structurally integrating a storage bunker into a building:

- woodchip bunkers with direct vehicular access
- conventional woodchip bunkers without vehicular access
- woodchip bunkers with preliminary drying
- woodchip bunkers with feed augers
- external underground woodchip bunkers
- external ground-level woodchip bunkers.

As all bunker variants can be operated for woodchips with the conveyor processes already described, the conveyor mechanism will not be described in detail here.

5.6.3.1 WOODCHIP BUNKERS WITH DIRECT VEHICULAR ACCESS

Woodchip bunkers with direct vehicular access (Figure 5.107) are spaces in buildings that open to the outside, and which can be driven into through a door. Thus the woodchips can be brought in with a front loader. To prevent the stored woodchips from sliding out of the bunker, the storage surface is enclosed with concrete or steel sleepers.

*Figure 5.107.
Woodchip bunker with direct vehicular access
Graphic: www.hargassner.at*

5.6.3.2 CONVENTIONAL WOODCHIP BUNKERS WITHOUT VEHICULAR ACCESS

Conventional woodchip bunkers without vehicular access (Figure 5.108) are situated within the building. They are separated from the rest of the space by a partition. The woodchips can be brought into this bunker with a conveyor belt or similar type of device. The woodchips are removed horizontally and fall through a chute to another auger. This then conveys the woodchips to the boiler for combustion.

*Figure 5.108.
Conventional woodchip bunker
Graphic: www.hargassner.at*

5.6.3.3 WOODCHIP BUNKERS WITH PRELIMINARY DRYING

With a woodchip bunker with preliminary drying (Figure 5.109), the fresh woodchips are stored on the warm ceiling of the boiler room so that they can be completely dried out. Once this has been completed, the woodchips are pushed manually or by machine over a parapet, where they fall into the actual bunker space. From here an auger transfers them to the boiler.

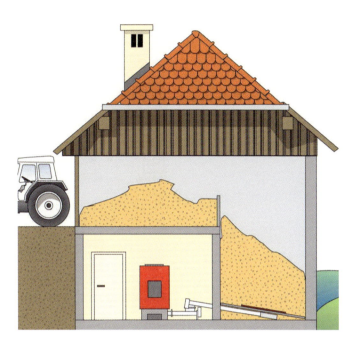

*Figure 5.109.
Woodchip bunker with
preliminary drying
Graphic: www.hargassner.at*

5.6.3.4 WOODCHIP BUNKERS WITH FEED AUGERS

In some cases boiler rooms and bunkers are located underneath buildings, making it impossible for vehicles to unload woodchips into them. Here heavy gauge screw conveyors can be used that transfer the woodchips from one storage bunker to the main bunker (Figure 5.110). These conveyors are permanently installed, and have diameters of at least 30 cm.

*Figure 5.110.
Woodchip bunker with auger extraction
Graphic: www.hargassner.at*

5.6.3.5 EXTERNAL UNDERGROUND WOODCHIP BUNKERS

Another possible variant when buildings have too little space for woodchip bunkers is to build external underground storage bunkers (Figure 5.111). These are built alongside basement boiler rooms and can be directly loaded by dumping woodchips from a lorry or trailer. These kinds of bunkers can also be easily filled manually or with conveyor belts and front loaders.

Figure 5.111.
External woodchip bunker
Graphic: www.hargassner.at

5.6.3.6 EXTERNAL GROUND-LEVEL WOODCHIP BUNKERS

If the woodchip boiler is installed in a ground-level boiler room, it is possible to construct an external ground-level store (Figure 5.112). This can be loaded through a feed hatch, with a front loader or with a conveyor belt.

Considerable experience is required when handling and transporting woodchips as a result of their difficult flow characteristics. Therefore woodchip bunkers should always be designed with the involvement of specialist designers and conveyor technicians who are experts in this area.

Designs made according to purely architectural criteria will generally lead to long-term problems with the operation. If it is planned to incorporate the bunker and its feed devices harmoniously within the building, experts should always be brought into the design process at an early stage. It is only through joint consultation that aesthetic mistakes or technical failures can be avoided.

Figure 5.112.
Ground-level woodchip bunker
Graphic: www.hargassner.at

6 Large-scale heaters

6.1 Introduction

Heating with wood in large buildings used by public institutions or private enterprises represents a big step towards sustainability.

Heating energy accounts for a third of the energy requirements of modern societies. If these energy requirements can be covered using local resources, this results in significant economic impacts on the region.

- The use of local natural resources creates independence, and reinforces local networking.
- The supply of fuel from the region creates an economic income.
- Large projects play a 'beacon' role, paving the way for other projects to begin.

Automatically fed heating systems are currently used as a way of utilizing biomass to supply energy to large buildings and estates. Woodchips and shavings are stored, then burned in these systems, with the resulting heat being directed to the connected consumers.

6.2 Implementing a wood energy project

Wood energy projects that involve a large heat station necessarily comprise a large number of economic, legal and technical elements. These are summarised in Table 6.1.

Table 6.1.
Technical, economic and legal aspects of large wood energy projects

Technical	Economic	Legal
Basic project conditions	Capital requirements	Organizational structure
Biomass volumes	Economic viability	Approval
Biomass provision concept	Financing options	Adjacent owners' acceptance
System design		Project planning

Because of the greater complexity in respect of logistics, operation and maintenance compared with fossil energy projects using oil and gas, the organizational structure in wood energy projects is of correspondingly greater importance. Depending on the type of the system being set up, the number of project partners, suppliers and operating personnel involved increases, and hence so too do the requirements for project organization.

The most important points for organizing and implementing a wood energy project are considered in the following sections.

6.2.1 Seven steps to a successful project

6.2.1.1 SELECT THE RIGHT SITE

It is important that the first project in a region is a 100% success. This concerns the economic efficiency, environmental sustainability, benefits to the region and the visual impression.

To this end, all existing buildings should be analysed and a pool of the best buildings selected for subsequent planning.

Advantageous conditions for a wood energy project are:

- a planned building that is going to be built in the near future
- a building that needs to be redeveloped and is to receive a new heating system
- an old heating system that is coming up for replacement.

Suitable buildings should have sufficient space available in the basement or outside the building for the fuel store and access for the delivery vehicles (HGVs). It may be possible to supply neighbouring buildings by creating a 'micro' local heat network.

6.2.1.2 CHECK THE AVAILABILITY OF FUEL
It should be established which fuels are available locally, as wood boilers cannot be fired with just any type of fuel. In particular it should be checked whether waste from woodworking industries or woodchips or pellets are available from local manufacturers.

6.2.1.3 SEEK PROFESSIONAL ADVICE
Large wood heating systems should be implemented only by experienced planners and technical experts. An inspection of the specified reference system on which the new system is to be based should form part of the preparatory programme.

6.2.1.4 INFORM THE AUTHORITIES AND GET THEM INVOLVED
The local authorities, political representatives and the public should be provided with full information about the project right from the outset. It is also important to show openly that people have had their right to a say in the decision-making process in order that a positive general sentiment will accompany the project. Also, wherever it is possible to do so, local workers and representatives should be integrated into the project.

6.2.1.5 SELECT A HIGH-GRADE BOILER
Nothing can do a project more harm than installing a boiler that smokes, is inefficient, and is difficult to operate. A quality product can be identified by the following criteria:

- efficiency > 85%
- low emissions: CO less than 200 mg/m^3, dust less than 150 mg/m^3 at full load and at 50% utilization
- automatic cleaning of the heat exchanger, and automatic ash removal
- possible remote monitoring capability for boiler parameters by manufacturer
- high reliability demonstrated in reference projects.

6.2.1.6 ENSURE THERE IS CAPABLE SUPPORT FOR THE SYSTEM IN OPERATION
A wood heating system requires competent monitoring at all times. There are two options available for this:

- An interested and qualified local government employee is responsible for the heating system. He or she takes care of purchasing fuel, checking the quality of the fuel supplied, and monitoring and documenting system operation, as well as cleaning the boiler and disposing of the ash.
- An energy company assumes full responsibility for the operation and maintenance of the system.

6.2.1.7 CELEBRATE THE PROJECT
A successful wood heating project is something special. This should be cause for celebration in the local community and in the region. Hold open days to help achieve high acceptance of the system.

Figure 6.1.
Wood heating station
Photo: Ingenieurbuero Gammel /
www.gammel.de

6.2.2 Basic conditions for local wood energy projects

6.2.2.1 CHECKLIST FOR COMMUNITY WOOD PROJECTS

6.2.2.1.1 Basic conditions

- No natural gas or local heat network in the local area — Yes = 0   No = 1
- Politically, the regional attitude is positive — Yes = 0   No = 1
- Local foresters are interested in supplying — Yes = 0   No = 1
- Local businesses are interested in supplying — Yes = 0   No = 1
- Total for basic conditions — =

6.2.2.1.2 Is sufficient fuel available?

- There are reserves from thinning the local forests — Yes = 0   No = 1
- Part of these reserves is already used for the production of chopped material — Yes = 0   No = 1
- Local farmers are interested in producing chopped product — Yes = 0   No = 1
- Chopped material can be sourced from cooperatives in the neighbourhood — Yes = 0   No = 1
- Dry residual products from timber processing are available — Yes = 0   No = 1
- Storage of chopped material is possible in existing buildings in the community — Yes = 0   No = 1
- Pellets are available — Yes = 0   No = 1
- Total for sufficient fuel — =

6.2.2.1.3 Are there buildings that are suitable for wood fuels?

- Buildings that have heating systems more than 15 years old — Yes = 0   No = 1
- Buildings that will need to be renovated in the near future — Yes = 0   No = 1
- Buildings with high heat requirements — Yes = 0   No = 1
- Buildings from which local heat networks can be supplied — Yes = 0   No = 1
- Buildings with sufficient space for boiler and fuel store — Yes = 0   No = 1
- Total of buildings suitable for wood fuels — =

6.2.2.1.4 Other favourable circumstances

- Activities in progress such as Agenda 21, Climate Alliance — Yes = 0   No = 1
- Existing initiatives for regional products — Yes = 0   No = 1
- Financial assistance for wood heating — Yes = 0   No = 1
- Positive experiences with wood fuels in neighbouring communities — Yes = 0   No = 1
- High level of interest in wood heating in private households — Yes = 0   No = 1

- Local manufacturers and traders interested in wood
  heating systems　　　　　　　　　　　　　　　　　　　Yes = 0　　No = 1
- Community has sufficient financial resources for an
  investment　　　　　　　　　　　　　　　　　　　　　Yes = 0　　No = 1
- Reliable contracting firms available　　　　　　　　　Yes = 0　　No = 1
- Capable and interested personnel to provide ongoing
  system support　　　　　　　　　　　　　　　　　　　Yes = 0　　No = 1
- Total of other favourable circumstances　　　　　　　=

6.2.2.1.5 EVALUATION

- Less than 10 points: Ideal conditions for a wood heating system. You should already be ideally placed for 100% renewable heat in public buildings.
- 11 to 20 points: The time is ripe for installing a first wood-fired boiler system. Begin by improving the basic conditions to enjoy successful project realization.
- More than 20 points: There is still a long way to go in the community. However, implementing even a small project would make a difference.

6.2.2.1.6 ADDITIONAL NOTE:
Differing results in the categories show where improvements can be made.

6.2.2.2 CHECKLIST FOR PRIVATE WOOD PROJECTS

Does the project make sense?

Is the developer ready?
- Does management support innovative projects?
- Does the company cater for customers who honour quality and technological progress?

Is a suitable project available?
- Is the planning about to begin or has it only just begun, so that it is easy to adapt?
- Is the project in an area that is not supplied with district heat?
- Is the project in a community that has an active environmental protection policy?
- In one of the developer's existing properties, does a heating system need to be replaced, and is sufficient storage space available?

Will energy services be offered?
- Examine local service providers and services offered.
- Check regional energy service providers.

Clarify fuel supply
- Find out who the suppliers are for wood fuels.
- Check existing supply structures.
- Would local forest owners be interested in supplying?.

Success factors:

A good boiler
- Efficiency > 85%.
- Emissions at full load less than 200 mg/m^3 CO and 150 mg/m^3 dust.
- Remote monitoring by the system manufacturer.
- Automatic cleaning of the heat exchanger and automatic ash removal.
- Possible remote monitoring capability for boiler parameters by manufacturer.
- High reliability demonstrated in reference projects.

Good planning
- Fuel storage of at least 30 m^3.
- Suitable access road for HGV deliveries.
- Boiler room approximately 20 m^3 and 2.5 m high.

*Figure 6.2.
Rotation boiler with ash removal
Photo: Koeb & Schaefer KG /
www.koeb-schaefer.com*

- Access to boiler room 120 cm wide at all points.
- Compliance with fire protection requirements.

Ensure residents and neighbours are well informed
- Supply residents with information about the function and advantages of the wood heating system.
- Inform neighbours about the planned heating system, selected boiler and its emissions characteristics, ideally with a visit to a reference site.
- Preliminary information to the local community and authorities.

Preparation of the operating phase
- Clarify who will supervise the boiler and dispose of ashes (caretaker, technician, residents).
- Training by the boiler manufacturer of the people who will be put in charge.
- Clarify ash disposal.
- Plan fuel delivery at times that will cause the least possible disturbance to residents.

*Figure 6.3.
Wood heating station with double boiler
Photo: Koeb & Schaefer KG /
www.koeb-schaefer.com*

6.3 Planning

6.3.1 Assessment of the project outline data

Heating systems exist with outputs ranging from 100 kW to 5 MW. These performance classes require large quantities of fuel, which means that stockpiling fuel and automatic feeding of the boilers are essential. It is also possible to use wood as a fuel in systems with an output in excess of 5 MW. However, these systems are mainly operated as combined heat and power plants, in which electricity is generated as well as heat. In terms of process technology, these systems are generally comparable to conventional condensing power stations.

The application areas for these large heating plants are generally municipal buildings or properties in the housing industry, as well as new residential estates with small and medium-sized local heat networks. Wood-fired heating systems can be designed to be fully automated. In addition they are built to have low maintenance requirements, with the result that they do not have to take second place to heating systems that use fossil fuels when it comes to emission values and plant management.

Figure 6.4.
Wood heating station
Photo: Ingenieurbuero Gammel /
www.gammel.de

6.3.1.2 BOILER OUTPUT

Wood boilers have a certain basic load, which means the boiler is optimally dimensioned if it attains at least 2000 full load hours per year. This generally ensures that 70% of the required energy for a building is provided by the planned wood-fired boiler. The remaining portion of the energy has to be supplied via a second, smaller boiler or other energy source.

6.3.1.3 STORAGE

The fuel stores for wood heating plants often need to hold considerable volumes of wood, according to the boiler output. The wood is mostly stored in silos or bunkers. The fuel is taken from these to be fed to the boiler using screw conveyors, hydraulic pushers or scraper chain conveyors.

The wood used for the combustion process must have already been dried before being brought to the fuel repository. Modern wood heating systems function only within the stated manual maintenance intervals if the use of air-dry wood with a maximum water content of 35% is ensured.

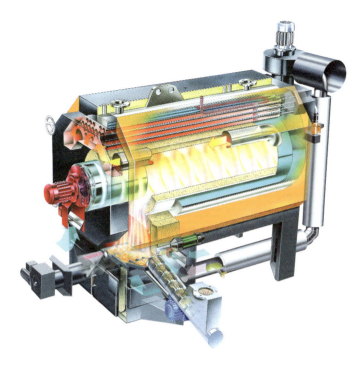

Figure 6.5.
Cross section of a rotation boiler
Graphic: Koeb & Schaefer KG /
www.koeb-schaefer.com

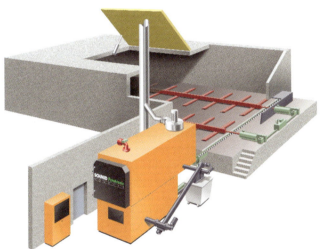

Figure 6.6.
Large-scale boiler with bunker
Graphic: Schmid AG /
www.holzfeuerung.ch

Figure 6.7.
Material bunker with hydraulic
push-floor
Photo: Polytechnik GmbH /
www.polytechnik.at

6.3.1.4 APPLICATIONS FOR MUNICIPAL BUILDINGS/HOUSING INDUSTRY PROPERTIES

A storage volume of 14 days operating at full load should be aimed for in the case of wood heating systems to supply residential buildings or municipal buildings. This is calculated using the following formula:

$$\frac{\text{Boiler output} \times 14 \text{ days' full load over } 24 \text{ hours}}{\text{Calorific value} \times \text{Bulk density} \times \text{System efficiency}}$$

For example:

- a boiler output of 1 MW
- 14 days' full load at 24 h/day
- an average calorific value of 4 kWh
- a bulk density of 250 kg per loose cubic metre
- a system efficiency of 85%

results in:

$$\frac{1000 \times 14 \times 24}{4 \times 250 \times 0.85} = 395 \text{ loose m}^3$$

It would therefore be necessary to hold a reserve of approximately 400 m³ storage volume.

6.3.1.5 APPLICATIONS FOR SMALL AND MID-SIZED LOCAL HEAT NETWORKS

Local heat networks of small and medium size can also be supplied by wood heating stations. In this case the heat generated in the boiler system is stored in water, which is pumped to consumers via an insulated copper, steel or plastic pipe system. At each node (connection point) in these networks there is a heat exchanger that transfers the heat from the central heating water circuit in the large system to the building's internal heating water circuit.

For the users of this type of local heat facility, a connection to a local heating network fuelled with wood energy means that they can benefit from the convenience of a fully automatic heating system without missing out on the environmental advantages of renewable energies. In addition, the owner of any such private home or apartment heated in this way is able to use spaces that would otherwise be occupied with heaters or fuel storage facilities for other purposes, as there is no need for an individual heating energy generation or fuel storage system.

In local heat networks the heat supplied to each individual consumer is recorded using heat meters. These are water meters that have temperature sensors in the flow and return. These devices calculate the amount of heat supplied each year from the volume of water passing through the meter and the temperature difference between the flow and return.

The following method can be used to estimate the annual consumption of this kind of heating plant:

$$\frac{\text{Nominal thermal output} \times \text{Annual full load hours}}{\text{Calorific value} \times \text{Bulk density} \times \text{System efficiency}}$$

For example:

- a boiler output of 1 MW
- 1500 full load hours per year
- an average calorific value of 4 kWh
- a bulk density of 250 kg per loose cubic metre
- a system efficiency of 85%

results in:

$$\frac{1000 \times 1500}{4 \times 250 \times 0.85} = 1765 \text{ loose m}^3$$

→ 1765 loose m³ × 0.25t/loose m³

440 t woodchips per year

The following method can be used for dimensioning the fuel stores for this kind of heating plant:

$$\frac{\text{Boiler output} \times 14 \text{ days' full load over 12 hours}}{\text{Calorific value} \times \text{Bulk density} \times \text{System efficiency}}$$

For example:

- a boiler output of 1 MW
- 14 days' full load at 12 h/day
- an average calorific value of 4 kWh
- a bulk density of 250 kg/loose cubic metre
- a system efficiency of 85%

results in:

$$\frac{1000 \times 14 \times 12}{4 \times 250 \times 0.85} = 197 \text{ loose m}^3$$

It would therefore be necessary to hold a reserve of approximately 200 m³ storage volume.

Large lorries are generally suitable in terms of the logistics for supplying both heating plant types. These can hold 80 m³ in the vehicle itself and 40 m³ in a trailer. The impacts of these logistics become clear especially for large systems with 6–8 MW, as these have a daily consumption of woodchips of up to three lorry-plus-trailer loads. The resultant traffic volume, noise pollution and harmful emissions have an impact on the immediate environment of this type of heating plant.

Figure 6.8.
External view of a modern fuel bunker
Photo: Polytechnik GmbH /
www.polytechnik.at

6.3.1.5 TYPES OF COMBUSTION SYSTEMS

There are many technical variations on the market for large heating systems fired with wood. These often have manufacturer-dependent advantages or disadvantages in respect of handling, maintenance intensity and the fuel quality requirements (wood type, water content and material purity).

Because of these factors, a planned wood heating project should always be implemented by specialists, who will analyse the basic conditions for a heating system including the fuel supply situation. Visits to reference systems and evaluation of the experiences learned from them are also important parts of selecting a boiler.

Table 6.2 gives an overview of automatically fed heating systems.

Table 6.2.
Automatically fed heating systems

Combustion type	Feed	Fuel	Power
Dual-chamber furnace	Mechanical	Woodchips, bark	35 kW – 3 MW
Underfeed furnace	Mechanical	Woodchips	20 kW – 2 MW
Stoker-fired furnace	Mechanical	Woodchips	200 kW and up
Cyclone furnace	Pneumatic		200 kW and up
Fluidized-bed combustion	Mechanical	Woodchips	10 MW and up

6.3.2 Assessing the economic efficiency

High investment costs are generally incurred in the development of wood energy projects in high power output classes. These costs are the deciding factor in assessing the economic efficiency of any such planned project. Any complete method for assessing the economic efficiency should always go through the steps listed in Table 6.3.

Table 6.3.
Steps for finding the energy generation costs

Calculation step	Parameters to ascertain
Determine basic project conditions	Location, energy demand data, heat requirements, possibly also the electrical power that can be used
Determine available fuel quantities	Regional biomass potential, average distance to biomass sources, seasonal changes, type of fuel and material properties
Logistic chain concept	Method of delivering fuels, type and duration of storage, further processing steps on site
Rough system design	Number, type and output of heat generators, boiler combustion type, flue gas treatment method, design data for system as a whole, machine data
Site engineering concept	Space requirements, buildings, pipe routeing, transfer structures

The project costs for wood energy projects comprise the following cost blocks:

Investment costs
- investment in system technology
- ancillary costs for planning
- incidental costs for approval, surveys
- taxes
- interest
- project reserves for problems.

Operating costs
- fuel costs
- personnel costs
- maintenance
- repairs
- insurance
- ash disposal
- equipment.

*Figure 6.9.
Wood boiler with push grill
Photo: Polytechnik GmbH /
www.polytechnik.at*

Incomes

The scale and type of combustion system can affect the income generated from the project. Income streams include:

- energy proceeds from heat
- energy proceeds from electricity
- financial assistance e.g. capital investment grants for promoting energy efficient and renewable energy technologies
- reduced interest rate loan.

*Table 6.4.
Typical investment costs for various wood energy projects*

	Machinery (%)	Construction (%)	Electrical (%)	Other (%)	Total costs (€)
500 kW boiler	70	15	3	12	150,000
1 MW boiler with building	55	30	5	10	300,000
5 MW heating station	55	25	10	10	1,200,000
10 MW steam boiler system with building	50	35	5	10	6,000,000
14 MW heating station	50	30	10	10	9,000,000

*Table 6.5.
Investment costs in local heat networks for new housing developments*

kW	€/m local heating network
200	200
500	225
1000	275
2000	300
4000	350

Typical investment costs for various wood energy projects are listed in Table 6.4. Investment costs in local heat networks for new housing developments are listed in Table 6.5.

*Figure 6.10.
Heating station with rotation boiler
Photo: Grafik: Koeb & Schaefer KG /
www.koeb-schaefer.com*

6.3.3 Fuel supply

In the context of organizing wood energy projects in high power classes, the supply of fuel plays an important role. Here the regional availability of fuels is essential. It is essential that the availability of potential fuel supplies is ascertained as part of preliminary investigations during the feasibility study. The logistical issues, distances and the seasonal availability of the fuels must also be determined.

The following aspects are critical when drawing up a fuel logistics concept:

- length of the transport routes
- seasonal availability of fuels during the heating period
- missing text
- form of delivery of fuels
- preparation of fuels.

To maximise local value creation, priority should be placed on a regional configuration for the fuel delivery contracts. The following supplier groups are relevant here:

- forest owners and forest producers' associations
- sawmills and wood-processing firms
- timber merchants
- forestry and horticultural operations
- recycling firms
- municipal facilities, timber yards, parks departments.

6.4 Legal organization

The type of energy facility being built has an important influence on the structure of a wood energy project. The organizational complexity of the project is also a decisive point. A wood combustion system for a single municipal building has a very simple project structure. This becomes even simpler if wood chippings from a regional supplier are used. By contrast, a system to provide local heat for a development area will be characterised by a large number of parties involved in the project (fuel suppliers, heat consumers, supplier firms) and therefore by a complex project structure.

Figure 6.11.
Material handling with loader
Photo: Polytechnik GmbH /
www.polytechnik.at

It is necessary to involve different parties in the project when organizing a wood energy project, according to the project's complexity:

- The owner or operator of the system is responsible for financing, building and operating the installation, and also for securing the fuel supply and – if applicable – the sale of the generated energy. The operator is a legal person. This can be an actual individual or a company.
- The system's plant manager can be a third-party institution (operating company, energy supply company) or the owner. This plant manager is responsible for maintenance, operation and also in part for the marketing of the generated energy.
- There are large differences between the various possible fuels that could be used. There is therefore a large range of possible fuel suppliers (e.g. foresters, local authorities, motorway maintenance authorities, sawmills).
- There is a need to distinguish between energy consumers for electricity and those for heat. The electricity generated in bioenergy systems is generally fed into the local grid operator's grid. This is a certainty in Germany owing to the fixed feed payment guaranteed under the Erneuerbare Energien Gesetz (Renewable Energy Act). There can be a wide variety of consumers for heat.
- Other parties in the project are generally the system suppliers. Their involvement depends on the type, size and complexity of the project.

6.4.1 Options for ownership arrangements

Unlike small combustion systems that are predominantly managed under private or company ownership, wood energy projects for large heating stations feature a large variety of possible ownership structures.

6.4.1.1 SYSTEM OPERATION BY LOCATION OWNER

This is the simplest form of operator model. Here the system is generally located on the same site on which the generated energy is used. In this case the investor and the operator are identical, as the system is operated by the location owner. The owner will also generally organize the fuel supply.

This concept can be expanded by also supplying neighbouring third-party properties from the existing location. In this case heat and possibly also power supply agreements with the neighbouring customers would need to be signed.

6.4.1.2 COOPERATION BETWEEN SYSTEM OPERATOR AND FUEL SUPPLIER

This kind of cooperative agreement is currently the most common way of operating an energy installation with small and medium-sized combustion systems. The operator is generally also identical with the project investor. The sale of heat and in some cases power as well takes place through long-term supply contracts.

Generally with this contractual arrangement with the fuel supplier, the supply contracts concern only certain qualities of wood. These are often supplied at fixed prices over an agreed period of at least one year. With smaller systems there is usually only one supplier. With larger systems, the supply risk for a large buyer is often spread across several suppliers who operate in the region.

6.4.1.3 LEASING AND HIRE PURCHASE

Leasing and hire purchase variations in the ownership structure of wood energy projects are longer-term agreements between the operators and the leasing companies. The leasing conditions, areas of responsibility and limits of supply are always specified on a project-by-project basis.

At the end of the contract period, with leasing agreements it is also possible for the operator to acquire the technical installations for a previously specified purchase price. Similar hire purchase options are in many cases offered by system manufacturers. These are always worth considering when the project operator's available liquidity is not sufficient to enable a direct purchase of the system.

The monthly payment amounts can be governed by a progressive or a linear lease payment agreement. In addition, the time of acquisition of the system at a defined residual value can be agreed individually according to the repayment amount and interest rate for the system. In a few cases, lease equipment rental agreements can be negotiated. These can often serve as transition solutions until completion of a project, for example. In such cases there is no fixed installation, the manufacturers instead providing container solutions. Here the contract customers organize the technical system operation under their own responsibility.

6.4.1.4 PROJECT FUNDS

With project funds, a particular project is financed by different commercial and private investors. Equity capital is provided for the project's financing and implementation. As well as a profit expectation, many investors are often motivated by the project's ecological approach. Regional identification with bioenergy projects and high acceptance are usually factors that increase investors' willingness to invest when they otherwise avoid speculative involvement on the stock market.

To prepare any such fund concept, the project's technical feasibility must be investigated. To ensure credibility with respect to investors, it is advisable to back the project up with recognised impartial engineers' reports. In addition, the economic viability of the project must be demonstrated in detail.

Important success factors for project funds are a precise description of the project (of the location in particular) and of the social and ecological synergies, and demonstration of a continuous, secure return on equity. In many cases finance agencies, specialized banks or private project companies will take on the administration of investment capital and professional marketing of the fund through their fund managers. Fund participation is generally represented via dormant equity holdings in the project companies that are set up, or as shares. The investor participates in the profits of the project company according to his participation level. If losses are incurred, the investor is liable with his paid-in assets up to the total loss of the money invested.

6.4.1.5 COOPERATIVE MODELS

In this form of organization, an operators' cooperative association is set up by several project participants with the aim of implementing the project together. Under company law, the cooperative is similar in structure to a society or association. It can therefore apply for non-profit-making status if necessary – for tax reasons, for example.

Figure 6.12.
Contracting boiler in a
portable container
Photo: Koeb & Schaefer KG /
www.koeb-schaefer.com

The cooperative association does not itself make profits; proceeds are distributed to the partners in the association. For project participants in agriculture and forestry, cooperatives offer the advantage of being able to have a presence as a producers' association or joint operation. This means there are many shoulders to support the risk of investment and operation. Other advantages of this form of cooperation are more efficient building use and machinery utilization and, also, the possibility of carrying out joint marketing.

6.4.1.6 CONTRACTING

Generally speaking, contracting describes models in which tasks in the area of financing, planning, installation, maintenance and repairs for systems that are part of technical building services are wholly or at least partly outsourced by the building owner to an external company. The term of this assignment is contractually agreed in each specific case.

With this kind of contracting it is also possible for compulsory energy-saving guarantees to form a part of the contracting model, as well as the energy-efficient and cost-effective supply of the useful energy that each specific customer requires. A typical contracting scenario involves contract durations of 10 to 20 years. During this time, the contracting company has complete responsibility for investment and for operating the systems.

The refinancing of the investments made by the contractor depends on the form of contracting. The refinancing can be generated from the sale of useful energy, margin benefits and the expenditure for system provision and management. In the case of housebuilding the supply costs can always be passed on to the users (tenants).

6.4.1.7 REGIONAL PARTNERSHIPS

Regional partnerships often prove to be a particular success model for the systematic widening of the use of bioenergy in regions. These models draw on economical, technical and logistical synergies between many individual regional partners.

Partners in regional partnerships can come from the waste disposal sector, farming and forestry, local authorities via municipal companies or timber yards. Investors or investor associations are often integrated into the partnership in order to secure the project financing. Potential plant operators may be considered if they have the necessary technical qualifications and sufficient practical experience. The final component in regional partnerships is suitable buyers of heat or other energy forms that may be produced such as cold and electricity.

7 Gasification

7.1 Introduction

Two hundred years ago biomass – mainly wood – was still our main source of energy. Yet without fossil fuels, the enormous population increases, the beginning of industrialization and the increasing standard of living would not have been possible. However, the annual global production of harvestable biomass is estimated at about five times current primary energy consumption. Thanks to huge progress in communications, transport and logistics, in the future it will be possible to exploit some of this potential commercially.

With sustainable agriculture and forestry, the CO_2-neutral biomass contribution to the global energy mix could be increased from its current level of 10% to a total of 20%, without harming priority foodstuffs production and industrial crop plantation. However, such a high percentage is possible only when, in addition to high-quality firewood for the production of energy, we tap into other residual agricultural biomass that is more difficult to utilize, such as sugar cane waste, corn silage, cereal straw, domestic animal litter and other organic waste sources.

7.1.1 Who should read this chapter?
Technically the gasification of biomass is currently still in the demonstration and market-entry stages. Nevertheless, in the medium term, gasification holds great potential for use in electricity production – particularly at decentralized, smaller bioenergy plants. Thus this chapter offers an introduction for those who are interested in biomass gasification and who wish to inform themselves on the status of the technology as well as on the environmental aspects and the economic possibilities.

7.1.2 What information is provided in this chapter?
This chapter introduces a tried-and-tested technology for generating electricity and heat from wood by means of gasification in an environmentally friendly way. In addition to providing general information, it will examine the status of the technology, utilization as energy, emissions and the economic viability of wood gasification plants. The organizational and planning aspects of wood gasification projects can be found in Chapter 9. Possible sources of support for bioenergy projects are outlined in Chapter 8.

7.2 System basics

Hitherto the most common and best-known form of using energy from biomass has been direct thermal conversion – that is, combustion. However, there are other ways of using solid biomass for energy to produce heat and electricity. One of these is gasification, in which solid biomass is converted into a combustible gas in a thermochemical process (Figure 7.1). The production of this secondary fuel has decisive advantages in terms of ease of handling and conversion possibilities for useful energy. In principle the same conversion processes take place as those used in combustion, but the different stages in the thermochemical conversion are separated physically and chronologically. This means the resulting product gas can be used in a combined heat and power (CHP) unit and utilizes the energy content of the fuel to maximum effect by means of combined heat and power.

Biomass gasification, and in particular the gasification of wood, is one of the most efficient and environmentally friendly possibilities for thermic utilization of biomass to generate electrical energy in small plants. Even in the period after the Second World War, gasification technology with wood gasifiers was available for commercial

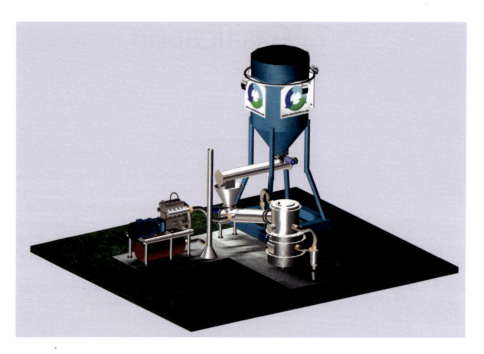

*Figure 7.1.
Model of a wood gasification unit
Graphic: Dobelmann /
www.sesolutions.de*

operation from the firm Imbert GmbH. It was forgotten in the years of cheap oil prices that followed, but now research is again being carried out in many places into the use of gasification technology, although so far there is no fully automatic, gasification equipment ready for the market. The main problem is the contamination by tar particles of the gas produced, which makes sustained use of this gas impossible in combustion engines. This problem can be solved in two ways: by improving the quality of the gas, and by developing innovative purification equipment. Today there are no more technical obstacles to the wide use of this elegant method of producing energy.

Thermochemical gasification of solid biomass is therefore an important technology of the future that can contribute to meeting the increasing demand for energy through combined heat and power in the coming decades.

The following sections describe the technique for gas production, the possibilities of using it as energy, project realization, and the economic aspects of biomass gasification.

7.3 Fundamental principles

7.3.1 Gasification

During the gasification process biomass is converted at high temperatures (over 600°C) into a new energy carrier in the form of a gas (i.e. product gas) as completely as possible. An oxygen-containing gasification medium (air for example) is applied to the heated biomass. The organic substances are broken down into combustible compounds, and the residual carbon undergoes partial combustion into carbon monoxide. The gasification takes place with sub-stoichiometric combustion ($0 < \lambda < 1$). The stoichiometric quantity of the oxidising agent is the calculated minimum amount to be applied to the fuel for complete combustion ($\lambda = 1$); the quantity of the oxidising agent is indicated via the fuel/air ratio λ.

The heat necessary for the process is generally supplied via partial combustion of the biomass used.

A fundamental feature of gasification is the physical and chronological separation of the production and utilization of the process product, gas. Herein lies the difference between combustion and firing. On this point Nussbaumer noted:

> When the gas produced is applied directly to a combustion chamber to produce heat, there is no thermodynamic difference between wood firing and wood gasification.

This illustrates that gasification is a sub-process of combustion, and that the gas is generated by partial combustion.

The low-calorie combustible gas produced, with an average calorific value of 5 MJ/m^3, can be used in burners for providing heat or in combustion engines or gas turbines for generating electricity or combined heat and electricity.

7.3.2 Fuel

The fuel properties are of critical importance in the selection of a gasifier. Different gasification models place specific demands on the fuels, such as pre-defined surface composition and moisture content. Long-term and reliable operation is possible only when the required parameters are observed. There are no gasifiers that can use all fuels and produce clean gas.

A reactor designed for the gasification of fist-sized wood briquettes, when used with wood chips, will produce less raw gas, plus a higher tar content in the raw gas, and will suffer other negative effects. Fuels that come in a wide range of piece sizes are usually not very suitable for gasification. Layers of such fuels have insufficient flow properties. They therefore tend to form undesired fragments, pits and cavities. Same-size wood pieces (especially cubes and spherical forms), on the other hand, are ideal.

In contrast to co-current gasifiers, counter-current gasifiers can also gasify fuels that do not have uniform surface properties. However, this advantage comes at a price: the gas has a very high tar content, and has to be purified with expensive filters in order to be compatible with an engine.

A gasifier performs optimally and has a good degree of efficiency only if the fuel intended for it is used and if there is optimal moisture content and fragmentation. Many gasifier manufacturers therefore set the gasifier to work with a specific fuel in the test phase.

7.3.3 Status of the technology

Wood gasification technology is currently used only in large plants. This is all the more surprising when one considers that the technology was already fully developed and widely used in Germany after the Second World War.

Although at that time no stationary plants for generating power and heat had been developed, engine technology for the automobile market was in place.

> Example: 29 MWel Amergas gasifier
> This biomass Amergas gasification plant in Getruidenberg, the Netherlands, is based on the circulating fluidized-bed gasifier, in which the product gas is also used in the steam generator of the connected coal power station to generate steam (see Figure 7.2). The plant has the capacity to gasify up to 150,000 t wood per year, which it can use to generate electricity with an efficiency of 35%, thus replacing coal as the primary fuel in the power station. Although the Amergas gasifier has so far been functioning without any problems, technical problems with the gas purification have led to time delays, with the result that the full annual fuel capacity has not yet been reached. The investment costs for the gasifier are about €1600/kW$_{el}$.

The Amergas plant is a fixed-bed gasifier. The biomass, usually fed in at the top of the reactor in solid fuel pieces, is exposed to the gasification medium and moves through various stages in a layer before reaching the ash pit.

In former times the principle of updraught gasification was often applied, in which the fuel and the gas move in opposite directions. This so-called countercurrent gasifier can be used today for plants with a fuel-heat capacity between 100 kW and 10 MW thanks to its design concept. Because of the high tar content in the gas, and the great demands this places on gas purification, a commercial breakthrough in combined heat and power cannot be counted on yet.

*Figure 7.2.
Large Amergas gasifier at the
Gertruidenberg power plant,
the Netherlands
Photo: Ecofys b.V. / www.ecofys.com*

With a co-current gasifier, downdraught gasification is applied. In other words the directions of the fuel and producer gas are the same. The gases decomposed in the pyrolysis zone are subsequently heated in an oxidation zone to over 1000°C. An extensive process of splitting the resulting long-chain organic compounds into short-chain compounds takes place, thus converting the tar-rich matter into low-tar matter. These react in the subsequent reduction zone with the ashes to form more gas (CO_2 reduced to CO).

This means that the output raw gas can be used where a high quality of gas is required. Co-current gasifiers are especially suited to combined heat and power generation for plants of low capacity (up to 500 kW).

In fluidized-bed gasifiers the flow rate of the gas is so high that a bed of material (usually quarry sand) from below circulates round the fuel. The fuel conversion and exchange of substances takes place spontaneously in stable combustion conditions and constant combustion temperatures, which ensures optimal combustion.

A distinction is made between a bubbling fluidized-bed gasifier, in which a clearly defined fluidized bed (height usually between 1 and 2 m) is characteristic, and the circulating fluidized-bed gasifier, in which the fluidized bed expands greatly.

These processes have undergone extensive testing, but have so far proved to be economically viable only for large plants due to the complex and expensive technology.

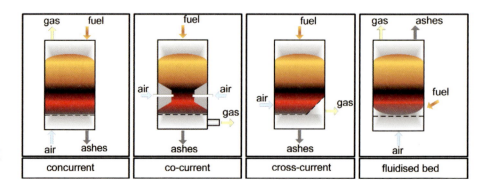

*Figure 7.3.
Types of gasifier
Graphic: Dobelmann /
www.sesolutions.de*

Figure 7.4.
Chicken manure gasifier, Bladel, the Netherlands
Photo: Ecofys b.V. / www.ecofys.com

Example: Chicken manure gasifier
The gasification plant in Bladel, the Netherlands, is an agricultural CHP demonstration plant for gasifying chicken litter with a capacity of between 60 kW_{el} and 40 kW_{el} (see Figure 7.4). The main incentive for this type of plant is the considerable costs that have to be paid in the Netherlands for disposing of chicken manure. The fluidized bed of this gasifier has a capacity of up to 900 t chicken manure per year. The plant will become operational at the end of 2003. The investment costs for this demonstration plant were approximately €8200/kW_{el}.

With entrained-flow reactors the gasification reactions take place during the pneumatic transport of the fuel through the reactor. The fuel must first be finely ground to make transportation possible and to ensure short reaction times for the gasification of the individual particles. Additional bed material, as needed for the fluidized-bed gasifiers, is not necessary.

Thus far this technology has not taken hold to the same extent as the fixed-bed and fluidized-bed gasifiers for the utilization of biomass because of the high costs.

Fixed-bed gasifiers using the co-current technique are suitable for commercial operation with decentralized solutions, especially in the area of combined heat and power; fluidized-bed gasifiers are more suited to larger plants.

The following sections will examine only the co-current gasification technique in more detail, as it is the most advanced in terms of economic efficiency and commercial status.

7.3.3.1 TECHNICAL SET-UP OF A CO-CURRENT WOOD GASIFIER

A wood gasification plant consists of a combination of various technical procedures. Figure 7.5 shows the structure of a co-current wood gasifier based on the Joos principle.

In the Joos gasifier, the fuel enters the gas reactor (3) through an input funnel (1) via a screw conveyer (2). Dust particles in the gas produced in the reactor are removed via a cyclone (4), and the gas is transported to the screw conveyer for indirect drying via a heat exchanger (5), before it can go on to be used as energy.

*Figure 7.5.
Structure of a wood gasifier based on
the Joos principle
Graphic: Dobelmann /
www.sesolutions.de*

7.3.3.2 GAS PRODUCTION FROM WOOD IN A CO-CURRENT FIXED-BED GASIFIER

For lower capacity plants – up to a maximum of 500 kW – fixed-bed gasifiers using the co-current principle are mostly used. The gasification zones of these gasification systems are illustrated in Figure 7.6.

The water contained in the fuel is first vaporised at a temperature between 100°C and 200°C (drying). The next stage is the degasification and thermic distillation of the contents into mainly gas elements at temperatures between 300°C and 600°C in the absence of oxygen (pyrolysis: $\lambda = 0$). Oxidation of the carbon and hydrogen takes place at temperatures usually over 600°C to cover the heat requirements of the endothermic reduction reaction and to break down the hydrocarbons that are formed in the pyrolysis zone. The wood gas is actually produced at temperatures of about 500°C by means of the reduction of the oxidation products CO_2 and H_2O of the glowing carbons. The basis for this is the Boudouard equilibrium of the carbon reaction and other equilibrium reactions, such as the water gas and methane equilibria, which are strongly influenced by the temperature and the pressure:

- Boudouard reaction:

$$C + CO_2 \leftrightarrow CO \qquad 162.2 \text{ kJ/mol}$$

- Hydrogen reaction:

$$C + H_2O \leftrightarrow CO + H_2 \qquad 119.0 \text{ kJ/mol}$$

- Methane reaction:

$$C + 2H_2 \leftrightarrow CH_4$$

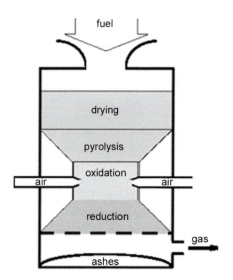

*Figure 7.6.
Gasification zones of a fixed-bed
co-current gasifier
Source: www.sesolutions.de*

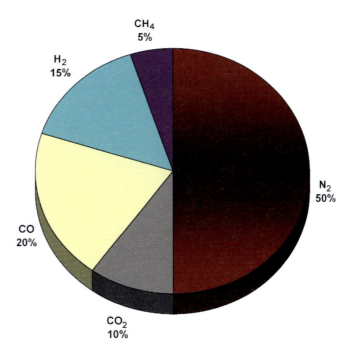

Figure 7.7.
Average composition of wood gas with air as the gasification medium
Graphic: Dobelmann / www.sesolutions.de

During the gasification process a gas is produced that consists of a mixture of combustible (H_2, CO, CH_4) and non-combustible (CO_2, N_2) gases. The average composition is shown in Figure 7.7.

The composition of the raw gas depends on the fuel characteristics (size of the pieces, moisture content and chemical composition), the gasification agent, the gasification temperature, and the pressure in the reactor.

7.4 Use as energy

7.4.1 Gasification applications

The gasification of biomass is a very promising technology – especially for generating electricity. The electricity provided has a high degree of efficiency. Furthermore, lower process-related emissions can be expected than is the case with electricity generation via direct biomass combustion. For that reason, much research has been carried out in the last few years to try and make this technology available for large-scale plants.

However, at the moment only a handful of gasification plants are operated commercially solely to produce heat. Plants for producing electricity – only here does the actual advantage of gasification fully take effect – currently exist only as pilot projects in the framework of research and development activities. There are difficulties particularly with gas purification, as the gasified biomass shows high dust content and sometimes considerable amounts of condensable organic materials.

Converted combustion engines and gas turbines, however, require an almost condensate- and dust-free combustible gas.

There is no ideal gasifier for the different types of biomass. The different gasifiers available have both advantages and disadvantages in terms of the biomass to be gasified, the desired gas quality, and the investment and operation costs. The different gasification systems differ in

- reactor type (fixed-bed, fluidized-bed/entrained-flow reactor)
- the method of providing heat (heat applied from outside or through partial oxidation of the fuel)
- the direction of flow of the biomass and gasification mediums (countercurrent or co-current gasification)
- the gasification medium used (air, oxygen, steam).

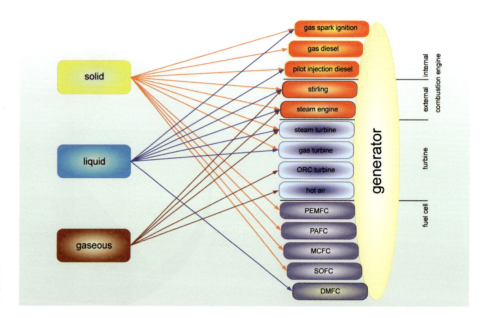

Figure 7.8.
Various procedures for generating electricity from biomass
Graphic: Dobelmann / www.sesolutions.de

7.4.2 Possible energy uses of gas generated from wood

The gas from biomass gasification can be used in several ways. It can be directly combusted, and the burnt gases produced can, for example, be used to generate heat or process heat or to fuel a thermal engine. However, the gas can also be used directly in a gas engine or gas turbine to produce methanol or hydrogen.

In future the Stirling engine and the fuel cell can also be a way of generating energy with generator gas.

Figure 7.8 provides an overview of the different procedures for generating electricity from biomass.

The burning of the gas in a gas engine produces almost 1 kWh of electricity per kg of wood and in CHP units generally twice as much thermal energy.

The following section discusses the advantages of combined heat and power.

7.4.3 Combined heat and power in a CHP unit

The most promising method of using produced gas from biomass gasification is in CHP plants. Chapter 2 looks more closely at the attributes as well as the advantages and disadvantages of combined heat and power: therefore this section will not examine these any further.

As a rule, industrial engines or car engines are modified and require conversion to be able to use gas from wood. Some problems specific to product gas arise. In particular, the condensate matter contained in the purified product gas can attach itself to the injection nozzles. These have to be cleaned up as part of the maintenance work. Deposits in the engine compartment shorten the intervals between oil changes, which have to be carried out approximately every 250 h of operation. The tolerance limits for tar compounds at which the use of product gas from biomass gasification in CHP units is still worthwhile are 100 mg/m^3 of product gas.

This means that converted diesel engines and gasoline engines based on diesel injection or extraneous ignition (spark plugs) are often used. Hitherto, the simplest and most successful method of operation has been with injection engines. Although these need 5–20% ignition oil to launch the combustion process, a consistent gas quality is not necessary. The absorbed product gas is enriched in the cylinder by means of direct fuel-injection with the quantity of diesel or biodiesel necessary for clean combustion. The exact amount of ignition oil that the engine needs to maintain the idle-running speed is injected into the engine. In the intake air a gas mixer is switched on, which then mixes the required amount of product gas into the air mass flow until the engine reaches the target performance. By regulating the quantity of ignition oil, it is possible to react to fluctuations in the quality of the product gas. When gas production has stopped completely, it is possible to run the engine on 100% ignition oil.

The use of gas engines makes more sense ecologically and economically, but it requires a higher quality of gas and is still in the trial phase as far as continuous operation is concerned. Many institutions and companies in Europe are still working on the development and launch of this technology.

As described in Chapter 2 in more detail, CHP units fired with the product gas can be operated in two different ways: with electricity, and with heat.

Concepts such as Stirling motors and the fuel cell are in the technical development stage, and are not yet suitable for use with product gas. Thus conversion of product gas as a fuel in a conventional small-scale CHP plant still faces no competition.

Combustion engines based on the principle of spark ignition or pilot injection diesel can look back on decades of practical experience.

7.5 Emissions and by-products

The by-products of gasifiers are ashes, condensate and sometimes carbon. In optimal operating conditions ashes can still have a carbon content of 25% (by weight). The condensate in non-contaminated wood consists mainly of water and low amounts of tar. Some gasifier manufacturers purify the ashes and condensate to the extent to which they are not considered hazardous waste. Usually the condensate can subsequently be put into the sewer system, and the ashes into a normal landfill site. However, some gasifiers produce a high concentration of toxic substances in their by-products (ashes, condensate), especially when using waste wood. These harmful by-products must then be disposed of at a cost, or the plant must be converted so that they no longer accumulate in this form. The carbon is extracted by some manufacturers and then has similar properties to activated carbon.

Other emissions are the fumes from the CHP units run on wood gas. These can occur in normal operating conditions of a CHP unit in two locations.

Emissions occur when the gas is flared and during normal CHP operations. Flaring is necessary in standard operations only for starting and turning off the CHP unit.

Major emissions in the raw gas are NO_x, CO, SO_2, and C_nH_m. As outlined above, the emissions differ according to the different types of gasifier. It has not yet been possible to assign definitive emissions regulations to CHP units operated with product gas from biomass gasification, but the tolerance limits to be observed will probably orientate themselves to those for solid mass firing.

The high CO content of the emissions requires intense purification in particular. In order to satisfy the emissions tolerance limits, the engine fumes must also be filtered with an oxidation catalyst.

The predominantly reducing conditions in the firing compartment represent one environmental advantage of gasification technology. No more than the quantity of oxygen needed to maintain the temperature for the subsequent pyrolysis process is supplied in the gasification reactor. This produces usable product gas during the conversion of the residual wood, but not nitrogen. The subsequent combustion of the gas in the engine of the CHP plant is controlled, with the result that combustion causes minimal toxic substances at every stage of operation. This is not always the case with conventional wood combustion plants.

In addition, the formation of other environmentally toxic substances such as dioxins and furan is prevented because of the low oxygen level of the surroundings. Overall, owing to the monitoring of reactions at every stage of the process, emissions are minimal, and discharge of harmful substances is regulated. Furthermore, the risk of the discharged ashes being burdened with harmful substances such as heavy metals is present only if the input materials are shown to have these contaminants as well. With the gasification of natural residual wood, from saw mills for example, it is still possible to use the resulting ashes as mulch.

7.6 Economic viability

The following analysis of economic viability serves as a guide for investment decisions. For an exact individual assessment, the marginal conditions must be adjusted as each case requires. The aim is to find out the maximum that wood gasification plants with differing capacities, at certain fuel prices and with electricity feed-in subsidies would cost in the future. Exact parameters and concrete cost figures for a full-scale gasification plant cannot be given, as the technology is still in the pilot stage and plants have until now almost all been built for test purposes.

7.6.1 Evaluation basis

The analysis of economic viability depends above all on the annual costs of a biomass gasification plant. These can be calculated from the total capital, consumption and operational costs. The calculation of these is explained in the following section. Note, however, too, that lifecycle costs are increasingly being focused on by end-users of all power generating mediums, and it is this factor that usually releases the funds for projects.

7.6.1.1 CAPITAL COSTS

The capital costs are calculated on the basis of the investment needed for the entire biomass gasification plant. In addition, the underlying interest rate and the amortization period have a decisive influence on the capital costs. To calculate the annual costs of the overall investment, a dynamic calculation model, the net present value method, is applied. This takes into account the timing of payments.

The net present value can determine the economic viability of an investment. The yardstick for assessing profitability with this method is consistent with the goal of maximising assets. The basis of the net present value method is to understand the funding of investments and projects as a series of payments.

With the net present value method all expenditure and income arising from the investment is discounted at the total discounting rate, and also discounted to the time directly before the initial outlay.

This means that the net present value method provides a very current value, which predicts how the value of the capital will rise or decrease in the future.

In simple terms, the capital invested is compared with assets with a bank. The interest rate is usually taken to be 8%. If the net present value is above zero, then the investment is more profitable than investing at a bank: in other words, the interest paid on the capital invested is higher than 8%.

The difference between annual revenue and expenditure gives the amount of the series of payments.

7.6.1.2 CONSUMPTION COSTS

These include the auxiliary energy costs, for example for pumps and screws, as well as the fuel costs. For example, a fixed purchase price for wood chips may be assumed, but this can be subject to fluctuations depending on supplier, quantity purchased, supply/demand and season. Higher fuel costs can be accepted when reliable, high-quality fuel is supplied, which increases the effectiveness of the plant, because a better quality of gas is produced. As a rule this also means repairs can be prevented.

7.6.1.3 OPERATING COSTS

The main operating costs are personnel and maintenance costs. The annual maintenance costs are usually assumed at a blanket rate of 10% of the investment costs. This also takes into account the shorter intervals between maintenance at some plants (especially changing the engine oil and fuel filter for CHP plants) or the use of more expensive operating resources (engine oil) sometimes stipulated. The costs for the personnel needed for monitoring and servicing the plant depend on whether the personnel already there are sufficient or whether additional staff have to be employed (dependent on size of plant and level of automation). Other costs (chimney sweep and administration, for example) are estimated as a percentage of the overall investment costs.

7.6.1.4 REVENUES AND CREDITS

Crucial for the economic viability of biomass gasification plants are the prices that can be charged or the credits that can be expected for electricity and/or heat. Most industrialized countries prescribe a minimum legal level of compensation for electricity produced from renewable energy and hence also from biomass. Chapter 8 contains an overview of the relevant assistance frameworks for bioenergy plants as well as advice on where to gain further information. However, the evaluation of the plant's own energy consumption of the electricity produced is based on the electricity prices applicable to different users: for rates-based customers (low voltage level) and customers with special contracts (mostly medium voltage level with high consumption).

7.6.1.5 FINANCIAL ASSISTANCE

The use of renewable energies and the operation of biomass gasification plants are supported by diverse financial programmes from public donors. More information can be found in Chapter 8.

7.6.2 Evaluation of economic viability

Gasifiers with smaller capacities (up to 100 kW) have not progressed beyond the demonstration phase. Consequently, no commercial products exist, which means that the investment costs for gasifiers are far higher than those for standard market products. Furthermore, at the moment only a rough estimate of the consumption and operational costs is possible owing to the lack of operational experience. The potential to reduce costs for gasification technology still needs to be exploited.

In the following example, profitability analyses were carried out based on operational experience gathered with a wood gasifier in order to estimate, amongst other things, the maximum investment costs for such a plant to still run profitably. The results of these analyses make it possible to conclude the maximum investment costs that such a system could bear. Changing the parameters of these analyses, for example the fuel costs and feed-in compensation rates for electricity, illustrates the direct relation of these marginal costs to the different parameters.

7.6.2.1 ENDURANCE TESTS FOR DETERMINING THE ECONOMIC VIABILITY OF A SMALL BIOMASS GASIFIER

Last year, on a farm in Bodnegg, several 200 h endurance tests were carried out on a co-current, fixed-bed wood gasifier (Joos) in cooperation with the Stuttgart Landesgewerbeamt (state trade authorities). A computer-aided performance analysis supplied the required data, in order to provide information and carry out an economic viability analysis. The basis for evaluating the economic viability was the energy yield from wood chips measured in the tests. On average, with 1 m^3 wood chips, about 200 kWh electricity (equals approximately 0.8 kWh per kg of fuel) could be generated in the tests. As the wood could be bought at a price of €8/m^3, the price for the fuel was €0.04/kWh of electricity generated.

For the evaluation of economic viability, a dynamic method of investment calculation was chosen, namely the net present value method.

The value is calculated from the revenues from the electricity and heat production, which have to be balanced with the investment and operating costs. The crucial factors in this regard are procurement of the biomass (assuming €8/m^3 of wood chips and €4/m^3 of waste wood from wood processing enterprises), the maintenance costs (about 10% of the investment costs), the renewal of the expendable parts (mainly sensors and engine parts), and the working hours (about 30 min/day).

Only a third of the heat produced was included in the calculation for the profitability analysis as the heating period was assumed to be from October to April, and because some of the heat has to be used for drying the biomass.

The following assumptions were used for the example profitability analysis:

- The interest rate is 8% in a perfect capital market.
- The series of payments lasts 10 years.

- The plant operates 6000 hours per year.
- Usable heat is 33% of the heat produced (only usable during the cold seasons, partly needed to dry fuel).
- Electricity feed-in compensation levels are 5 cents, 10 cents or 15 cents per kWh.
- The heat generated has a value of 4 cents per kWh (dependent on oil prices).
- Maintenance costs are about 10% of the investment total.
- Working hours per day (220 working days) equal 30 min and are calculated to cost €15/h.

As neither the investment costs for wood gasification plants, including the feeder and gas purification equipment, nor the price for a CHP plant were known, the maximum possible investment costs were determined for the results from the endurance tests. According to these, for plant sizes of 30 kW electrical capacity, the maximum incremental investment costs were between €50,000 and €60,000, depending on the wood prices. In today's general conditions, smaller micro wood gasification projects would be profitable from an economic perspective only in certain circumstances.

Adjusting the parameters to CHP units with a greater capacity and varying the wood price provides the results shown in Figure 7.9.

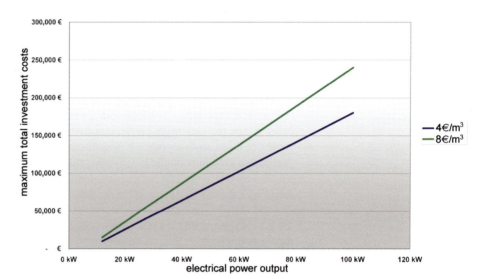

Figure 7.9.
Maximum total investment costs for small-scale wood gasifiers
Graphic: Boettger / www.sesolutions.de

Thus the maximum marginal investment costs for a biomass gasification plant depend particularly on capacity, electricity subsidies and the price of the fuel.

As with all bioenergy projects, the fuel logistics expenditure increases with the size of the plant, and it becomes more difficult to use the resulting heat to full capacity.

In Figure 7.10 the feed-in compensation was varied and the wood price was set at €8/m^3 of wood chips. With feed-in compensation below 10 cents per kWh of electricity produced, operation of the plant would be unprofitable, as the fuel costs would amount to more than half of the feed-in subsidies. At 10 cents the maximum investment costs are between €1620/kW (plant capacity 30 kW$_{el}$) and €1890/kW (plant capacity 100 kW$_{el}$). At 15 cents compensation the maximum investment costs are €2800/kW (plant capacity 30 kW$_{el}$) to €3100/kW (plant capacity 100 kW$_{el}$), and with feed-in compensation of 20 cents, up to €4250/kW would need to be invested for a 100 kW$_{el}$ plant.

The analysis of economic viability showed that the gasification of biomass at the lower end of the capacity range can be a profitable procedure for generating electricity and heat, as well as for disposing of biowaste.

In the endurance tests the wood gasification plant, with balanced combined heat and power in a CHP unit and based on the assumptions made, could cover the investment and running costs with the heat and power yield and generate a profit.

Because of the estimated doubling of energy needs in the next 50–60 years and the significant reduction in the use of fossil fuels, the use of energy in an intelligent,

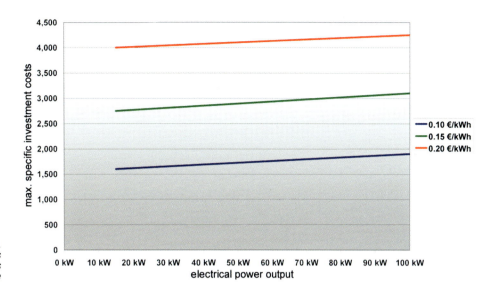

Figure 7.10.
Maximum specific investment costs
for small-scale wood gasifiers
Graphic: Boettger / www.sesolutions.de

effective and consequently decentralized manner will become increasingly appealing. But it will also be necessary, in order to meet the growing demand for energy, for other fuels and technologies to play a greater role in meeting additional energy requirements.

Wood gasification technology is a promising alternative for helping to meet these additional energy requirements. To achieve this in the next few years, the technology must be developed for commercial operation and the system and operating costs must be reduced.

8 Legal boundary conditions for bioenergy systems

This chapter provides the reader with an introduction to the relevant legal issues related to the erection and operation of a bioenergy system. It considers the general aspects valid for any bioenergy system and system-specific details for the different technologies, as there are anaerobic digesters, biofuel-driven energy applications and solid-biomass-fuelled combustion units.

As the regulations are constantly changing the chapter does not provide details of specific acts and ordinances, but merely names the relevant aspects that commonly influence bioenergy projects. Supporting this is a list of references to the most important sources of detailed information.

8.1 Introduction

If we look at the legal framework that is relevant for the installation and operation of bioenergy plants, we can distinguish three areas:

- general legislation relating to the installation and operation of renewable energy systems and their connection to the electrical grid
- system-specific legislation relating to the erection and operation of particular bioenergy systems (biogas plant, biomass combustion unit providing heat and/or power, or biofuel-driven combined heat and power (CHP) engine)
- specific legislation relating to the biomass input to the plant.

8.1.1 General legal aspects

The first area covers the general framework for feeding electricity into the electrical grid. In the past the feed-in of electricity from independent power producers to the electrical grid needed to be negotiated for each new power plant with the respective grid operator. Today, in most developed countries this aspect is commonly regulated by special schemes for electricity from renewable energy sources or from CHP engines or plants. These cover specific issues such as preferred grid access at reasonable cost. Usually an appointed governmental body – often part of the ministry of economic affairs or of industry and energy matters – gives the necessary authorization.

8.1.2 Erection and operation of bioenergy systems

In general, bioenergy systems require several approvals and authorizations before they can be erected and operated, and before heat or power can be supplied. Usually the approvals procedure for bioenergy systems is twofold, depending on the characteristics of the plant. On the one hand a building permit needs to be obtained to allow the erection of a new structure, and on the other hand compliance with the relevant national emission regulations needs to be proven. For special cases, and particularly for larger plants, additional environmental impact assessments are required, in order to demonstrate that the environment around the particular system will not be damaged, as laid down in the relevant environmental regulations. In contrast to this, small-scale heating systems such as open fireplaces or pellet ovens often require no permits, just regular emissions measurements commonly carried out by chimney sweeps.

In general the legal framework relating to the erection and operation of a bioenergy system consists of legal acts and ordinances that have to be obeyed. In addition, detailed regulations explain the procedures to be followed in order to obtain the required permits for the particular bioenergy system.

In addition, during the technical execution of the installation certain technical rules and standards need to be followed by the installing companies (e.g. technical rules on electrical installations), which are usually laid down by craft associations, standards institutes or similar bodies. However, registered installation companies are usually liable for their work according to the applicable technical rules.

8.1.3 Biomass-related legal issues

The wide range of biomass, its diverse characteristics, the different origins and the various types of application show the complexity of setting up bioenergy projects. For instance, looking at one segment of solid biomass – woody biomass – the range regarding the quality of wood extends from fresh forestry residues to highly contaminated waste wood. Input material for anaerobic digestion plants shows an even larger variety, including manure, residues from the food industry including meat production, the organic fraction of household waste and energy crops. Each biomass stream has to be treated in a particular way so that it can be employed as a fuel to generate electricity and heat. But alongside the technical aspects there are also legal issues to be considered for each biomass stream. In general the different biomass streams are categorised for three main reasons:

- the eligibility of a particular material as a renewable energy carrier, which is also important in order to make use of support measures
- classification of a specific biomass and the related emissions regulations
- identification of the need for special treatment of an input or output material.

8.2 General approval issues for renewable energy systems

8.2.1 Grid access permits

When planning to set up a bioenergy project in which electricity is generated and fed into the grid, it is essential to apply to the designated authority for grid access. Commonly this authority is either a government body or the grid operating company. In most countries this is only a formality provided the system complies with the relevant technical standards. However, certain circumstances may affect the costs related to grid access, apart from the normal approval fees. For example, the capacity of the local electricity lines or converter substations may be smaller than required, so that additional line capacity needs to be built up, or the system has to be connected to a more distant line with larger capacity. In particular, with wind energy systems the issue of the compensation of reactive power may add to the grid connection costs.

8.2.2 Building permits

Commonly bioenergy systems have to be granted building permits for erection of the system and the related buildings. In some countries where a number of renewable energy systems have already been erected there are special building codes for these systems. However, in many countries such systems are viewed as energy conversion systems – similar to any other power plant – and related buildings. Sometimes this classification can bring up requirements that go beyond those that are necessary for a bioenergy project. Therefore at an early stage of a project it is advisable to check with the respective authority which particular regulations have to be complied with. Usually it is the thermal capacity of the bioenergy system that determines whether a local, a regional or even a federal authority is empowered to approve the installation. In cases where local authorities are involved, it may be the first time that the authority has been confronted with such a system, and therefore the approvals process may take longer than expected.

8.2.3 Technical requirements

In the process of the erection of technical systems in general, various technical skills are necessary that only authorised craftsmen can provide. For example, there are common rules for the installation of electrical equipment. Specifically, when working on the grid connection of systems, only electricians who are registered with the particular grid operating company are allowed to perform this work.

Several steps also require authorised inspection agencies to check whether an installation has been carried out according to the valid technical regulations and standards. In particular, safety systems and pressurized vessels are subject to such authorizations.

8.3 The approvals process for bioenergy systems

Various regulations need to be complied with in order to receive a permit to build and operate a bioenergy plant. For smaller systems such as small-scale biomass heating (e.g. pellet ovens) the required permits are mostly limited to compliance with current emissions regulations and certain safety rules on the handling of fuel. Local municipal authorities usually deal with these.

For larger bioenergy systems, the approvals process is becoming increasingly complex, and several authorities – not just municipal ones – are involved in granting the required permits.

During the entire project development phase and the related approvals process it is important to contact the respective authorities at an early stage. They should be able to provide the project developer with the necessary details of exactly what information is needed in order to apply for the required permits, and by contacting them early and on a regular basis the developer may save time and money during the approvals process. Furthermore, a good relationship with the approving authorities is often essential for the viability and the success of a bioenergy project.

In general, the approvals procedures differ from anaerobic digestion plants to solid-biomass-fuelled combustion systems providing heat and/or power to stationary heat and power plants supplied with liquid biofuels. The approvals are usually related to the following areas:

- fuel (type and characteristics of the biomass used)
- emissions
- residues and waste.

An overview of the legal framework and related approvals issues for combustion systems and anaerobic digestion plants is given in Figure 8.1. It shows the four main categories of permission, which are related to the input (biomass), the output (residues), general issues (buildings and grid-connection) and aspects of permission for combustion systems in general. The last category would also apply for any other combustion system using any type of fuel.

8.3.1 Biomass input

A country's legislation normally provides rules on which energy sources are considered as renewable energy carriers and are therefore eligible for national support measures. There are usually further distinctions defining the type of biomass, including the source of each biomass stream and the technologies that are allowed to convert the biomass into heat and electricity. In addition certain requirements regarding pre-treatments of biomass are sometimes given. The main reasons for such detailed definitions lie in the diverse and manifold types of biomass stream and the need to comply with the respective legislation. For example:

- In some countries waste is viewed as a renewable fuel, whereas in others it is not.
- Peat is considered as biomass in Finland, but in most other countries it is classified as a fossil fuel.

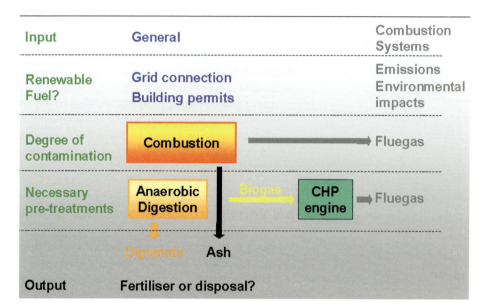

Figure 8.1.
The legal framework around a bioenergy project

- The degree of contamination of waste wood determines whether or not it is categorised as a renewable fuel. For example, in Germany a specified limit of PCB/PCT or dioxins/furans in waste wood may not be exceeded, otherwise it is considered as special waste, which needs to be removed according to special rules.
- Animal residues from meat production or the organic fraction of household waste generally need to be sanitised before being allowed as digestible material for a biogas plant.
- In the Netherlands, electricity that is produced by co-combusting biomass in large coal power plants is eligible for support measures for renewable energy sources.
- In most countries the type of input material in an anaerobic digestion plant determines whether the digestate can be used as a fertilizer or needs to be removed.

Because of the large variety of biomass streams, and widely varying technical applications required to make use of the biomass as an energy carrier, in the following the distinction is merely made between the two major categories of application: anaerobic digestion and combustion systems. As stationary biofuel applications consist of either a CHP engine or a boiler the relevant legal aspects are treated either in the anaerobic digestion part (CHP engine) or in the combustion system part (boiler).

8.3.2 Emissions

In general, the term 'emissions' refers to emissions of particles or gases that are harmful to the environment and human beings. As well as these emissions, noise also has to be considered when operating a technical installation. However, noise emission regulations are generally less important – and easier to comply with – than gas and particle emissions during an approvals process.

Wherever a fuel is incinerated in a combustion process – a boiler, an oven or the combustion chamber of a CHP engine – flue gas is emitted. Such gas commonly consists of a mixture of different gases and particles. The main elements are oxidised components of the fuel, such as carbon dioxide, carbon monoxide, nitrogen oxides and sulphur oxides, plus heavy metals and non-oxidised particles of the fuel – commonly referred to as dust. Other components can be substances that are formed by the influence of high temperatures and pressures during the combustion process, such as hydrochloric acid, or even dioxins when incinerating certain fractions of contaminated waste wood.

Compliance with the emissions regulations is one of the central elements that needs to be proven during the approvals process. A granted permit commonly allows only those fuels that are dealt with in the permit documents. Because of the different emissions that originate from distinct fuels, any change of fuel requires a new permit.

Hence, once a bioenergy plant is built and operating, great care needs to be taken that only the approved fuel is used.

In most country's emission regulations, differences are made depending on the type of fuel used, the power capacity of the combustion process, and the type of technology employed. For example, different emission limits can be found according to whether wood or straw is incinerated or whether the combustion power capacity is 100 kW or 20 MW.

The set emission limits in the relevant regulations generally call for certain flue gas cleaning technologies, measurement equipment and control devices. This obviously influences the investment costs and thus the feasibility of the project.

8.3.3 Technology-specific aspects

In the following, technology-specific issues related to the legal framework and approvals are presented, distinguishing between anaerobic digestion, small-scale heating systems, and larger combustion systems.

8.3.3.1 ANAEROBIC DIGESTION

A distinctive aspect of biogas plants is the large variety of potential co-substrates that can be used as input material for an anaerobic digester. This variety may lead to a set of different requirements regarding pre-treatment, and restricting the usage of the digested material as fertilizer.

There are usually regulations that determine particular rules of treatment for specific biomass streams, particularly waste such as residues from meat production or the organic fraction of household waste. This often implies that permits need to be acquired for each co-substrate.

For agricultural biogas plants it is important that the digestate can be used as fertilizer on agricultural land rather than being treated as waste, which might need to be removed at high costs. Therefore great care needs to be taken when deciding upon the addition of certain co-substrates. Some co-substrates come with higher levels of heavy metals or other components that may contaminate the digestate and thus prevent it from being spread on the agricultural land.

In addition to the legal aspects related to biomass, the CHP engine as a combustion engine requires a permit too. Here, emission avoidance or reduction plays the key role. In addition, the handling of gas calls for compliance with the relevant safety rules for the operation of the CHP engine.

8.3.3.2 SMALL-SCALE HEATING SYSTEMS

Common biomass heating systems are manufactured in series production and come with a certificate of compliance with the relevant standards and norms, including the emission regulations for the particular heating capacity. Therefore in most cases no specific permits need to be acquired, but regular emissions control measurements – usually carried out by the chimney sweep – have to be proven.

In addition, certain safety precautions need to be followed when storing biomass as feedstock for the heating in order to avoid accidents and minimize the risk of fire.

8.3.3.3 LARGER COMBUSTION SYSTEMS

One of the most important steps during the development of a larger bioenergy project is early contact with the approving authorities. These authorities provide the project developer with the necessary information about the required documents and expert reports, as well as the current approvals practice for the particular system type.

As for biogas plants, the approvals process for larger combustion systems centres on compliance with the emissions regulations. As noted in section 8.3.2, the emissions limits depend on the power capacity of the particular plant and the type of bioenergy carrier used.

In contrast to small-scale heating systems and the smaller biogas plants, approvals for larger combustion plants may also include environmental impact assessment studies and an involvement of the public. Such requirements on the plant operator are usually defined in the relevant acts and ordinances. In general, the more contaminated a fuel is and the larger the plant, the more stringent are the requirements during the approvals process.

8.3.4 Documents accompanying the approvals process

Various documents need to be provided to the different approving authorities accompanying the formal permit application forms. They include the following:

- description of the project
- general drawings
- description of the biomass
- process flowcharts
- operating times
- technical data on system and components
- emissions reduction measures
- safety measures
- residue and waste removal.

Some information to be provided to the approving authorities needs to be expert reports prepared by external independent and officially accredited organizations.

8.4 Further information

To obtain up-to-date information on the specific legal framework it is advisable to contact the relevant authorities in the country where a bioenergy project is being developed. The authorities, and the approvals they are responsible for, differ from country to country, so in the following the relevant authorities are named for a selection of countries. Sources of additional information are also listed.

8.4.1 UK

8.4.1.1 APPROVING AUTHORITIES IN THE UK

Grid access permits:

Office of Gas and Electricity Markets (regulator)
www.ofgem.gov.uk
CCL exemptions and other electricity licensing information.

Department of Trade and Industry
www.dti.gov.uk
The DTI is responsible for planning consent for grid connection in England and Wales; for Scotland contact the Scottish Executive: www.scotland.gov.uk.

Local electricity company
You may need to contact your local electricity company about linking into the local distribution network, although it is advisable to contact the DTI first, as the electricity market is now deregulated.

Building permits:

Local county council and district council planning departments
(Website depends on location of proposed site)
Building proposals for planning permission are first submitted locally and must conform to local development plans. Local planning authorities will provide permits for power plants < 50 MW.

Department of Trade and Industry
www.dti.gov.uk
This department is responsible for approvals of power plants >50 MW.

Association of National Parks' Authorities
www.anpa.gov.uk
If development is on a national park, the relevant authority will have a role in the planning application.

Office of the Deputy Prime Minister
www.odpm.gov.uk
Details of planning guidance, e.g. PPG 22 on renewables, that will affect development's approval.

Environmental permits:

The Environment Agency
www.environment-agency.gov.uk
The Business section in particular provides information about permits and licences for compliance with major environmental legislation in the UK. Includes air emissions and waste permits.
 Note: If an environmental impact assessment is needed, a wide range of stakeholders will need to be informed, along with the Environment Agency, including the local planning authority, local council (if different) and the Countryside Agency.

8.4.1.2 ADDITIONAL INFORMATION

Renewable Power Association
www.r-p-a.org.uk
Information about the renewables arena in the UK, including legislative issues, with dedicated members' area.
Department of Trade and Industry
www.dti.gov.uk/energy/leg_and_reg/index.shtml
General legislation on power stations and planning consent, with contacts.

Department of the Environment, Farming and Rural Affairs (DEFRA)
www.defra.gov.uk
Provides the latest information on developments in environmental legislation, such as waste and air quality standards.

British Bio Gen
www.britishbiogen.co.uk
Trade body for bio-energy in the UK; provides detailed information to members about permits and licensing.

8.4.2 USA

8.4.2.1 APPROVING AUTHORITIES IN THE USA

Grid access permits:

Federal Energy Regulatory Commission
www.ferc.gov
In charge of regulation of interstate transmission of electricity; also contains information on regional transmission organizations (RTOs) and interconnection for both small and large generators. Should have contacts in relevant states and with relevant transmission organizations.

Great Lakes Regional Biomass Energy Program
www.cglg.org/1projects/biomass/index_frame.html
Provides information on all permits, divided by technology type, as well as other helpful information covering seven states.

Building permits:

Local/municipal authority
First contact the local planning authority/city council or zoning board. They will provide information about the planning processes.

State authority – often environmental protection or energy authority as lead authority
May be additional to local authority permit, or could supersede it.
Federal authority, e.g. Bureau of Land Management
www.blm.gov

United States Forest Service
www.fs.fed.us
Sometimes federal authorities will also need to be involved; this depends on location.

California Energy Commission
www.energy.ca.gov/siting/index.html
For California only – comprehensive approvals procedure for power plants >50 MW. Similar energy commissions may also exist in some other states.

Environmental permits:

Environmental Protection Agency
www.epa.gov
National organization that sets standards, then delegates approval responsibility to states and tribes. Contact EPA for information about the relevant state.

8.4.2.2 ADDITIONAL INFORMATION
Regional Biomass Energy Programme (RBEP)
www.ott.doe.gov/rbep/
Information about five regional bioenergy programmes administered by the Fuels Development Office within the DoE's Office of Transportation Technology. Research, but also aims to help support investment into bioenergy roll-out.

8.4.3 Canada

8.4.3.1 APPROVING AUTHORITIES IN CANADA
Grid access permits:

Ontario Energy Board
www.oeb.gov.on.ca

Alberta Energy and Utilities Board
www.eub.gov.ab.ca

Ministry of Energy and Mines, British Columbia
www.em.gov.bc.ca

Saskatchewan Industry and Resources
www.ir.gov.sk.ca

Electricity is regulated by an Energy Board in each province, which, along with the Federal government, is responsible for the rules in its jurisdiction. Contacts here may not cover every province, but simply look up the Energy Board or provincial energy ministry

Building permits:

It is best to contact the city/municipal government that covers the area of the proposed site, as well as finding out the provincial standards.
   Building permission is controlled at a local level, to be in keeping with provincial standards.

Environmental permits:

Environment Canada – energy and electricity section
www.ec.gc.ca/energ/electric/elec_home_e.htm
National agency for environmental protection and weather forecasts. Provides up-to-date information on all environmental legislation. Contact them for approvals advice, especially as different bodies may operate province by province.

8.4.4 Australia

8.4.4.1 APPROVING AUTHORITIES IN AUSTRALIA
Grid access permits (need to apply at the state level):

Essential Services Commission of Southern Australia
www.saiir.sa.gov.au
Licensing for generators in Southern Australia; information on transmission organizations.

Office of Energy in Queensland
www.energy.qld.gov.au
Powerlink
www.powerlink.com.au
Information about the electricity market in Queensland; links to the transmission company (Powerlink).

Essential Services Commission Victoria
www.esc.vic.gov.au
Provides licences to generators (and others in the electricity industry) and provides other information to these sectors.

Independent Competition and Regulatory Commission (ACT)
www.icrc.act.gov.au
Licensing information and information about other licensees in the industry in Australian Capital Territory.

Ministry of Energy and Utilities (NSW)
www.doe.nsw.gov.au
Information on electricity and energy in New South Wales.

National Electricity Market Management Company
www.nemmco.com.au
Good information and links available; central website for the National Electricity Market Managers.

Office of the Renewable Energy Regulator
www.orer.gov.au
Apply here for accreditation as a renewable energy source.

Australian Competition and Consumer Association
www.accc.gov.au
National body responsible for regulating the transmission service providers in the National Electricity Market. Will have some information available.

Local electricity utility
They may also be able to provide advice on feed-in generators, particularly where these are small scale.

Building permits:

Local planning department and then state level
For most information contact the local council town planning department first, although it is likely that large projects will have to be discussed with the state authority as well.

Australian Building Codes Board
www.abcb.gov.au
Information on all building regulations and codes.

Environmental permits:

Department of the Environment and Heritage
www.ea.gov.au
Details of permits and approvals administered by the Ministry.

Office of the Technical Regulator, Southern Australia
www.energyandsafety.sa.gov.au
Responsible for monitoring safety and technical issues for generators, and transmission and distribution entities.
Office of the Chief Electrical Inspector (VIC)
www.ocei.vic.gov.au
Responsible for safety issues in the electrical industry.

8.4.5 Scandinavia

8.4.5.1 APPROVING AUTHORITIES IN SCANDINAVIA
Grid access permits:

Danish Energy Authority
www.ens.dk
Information and contact point on all energy-related policies.

Fingrid Oyj
www.fingrid.fi
Finnish grid operator – obliged to permit any producer access to grid. Contact them for fees and procedural information.

Finnish Energy Markets Authority
www.energiamarkkinavirasto.fi
Ensures fair pricing in the transmission and distribution of electricity. Settles disputes.

Swedish Energy Agency
www.stem.se
Ensures fair pricing in the transmission and distribution of electricity by regulating the monopoly transmission system.

Statnett – Norwegian grid operator
www.statnett.no
Operates the transmission grid. Contact for fees and procedural information.

Norwegian Water Resources and Energy Directorate, Ministry of Petroleum and Energy, Norway
www.nve.no
Responsible for regulating non-fossil energy, transmission etc. in Norway. Will have information on process for new generator installation.

Note: In all cases, for small developments, connection to the national high voltage grid may not be necessary, and the best contact may be made with the local distribution network in the form of the local electricity company.

Building permits:

Local/municipal planning authority (DK, FI, SE, NO)
It is essential to obtain local planning permission. This will also ensure that the development adheres to any energy plans that are part of the local development plan. The authority will also usually be able to advise on building codes.

Swedish Energy Agency
www.stem.se
In Sweden the Energy Agency is the central authority for local planning relating to energy, and so approvals may end up here. Also responsible for granting permission for new transmission lines.

Ministry of Trade and Industry, Finland
www.ktm.fi
Responsible for land use and planning issues for some aspects of electricity, such as power lines.
Ministry of Petroleum and Energy, Norway
http://odin.dep.no/oed/engelsk/index-b-n-a.html
This ministry can have the final say on large energy-related developments

Environmental permits:

Danish Environmental Protection Agency
www.mst.dk
Information on Danish policies and legislation to do with air quality and climate change as well as other environmental issues. Contact them for more information on approvals procedures.

Local authority environmental department or county environment department (Denmark)
The approving authority in Denmark is either local or county-based depending on the size and nature of the proposed generator.

Environmental Permitting Authorities, Finland (part of the Ministry for the Environment)
www.vyh.fi
Responsible for approving heavily polluting activity. Three divisions according to geographic region. Contact through environment agency to find out how to procure the correct permits. Enquirers may be referred to a Regional Environmental Centre, which is an even more localised service.

Danish Energy Agency
www.ens.dk
For some environmental requirements, they will be responsible for approvals.

Swedish Environmental Protection Agency
www.internat.environ.se
Approval is likely to be the responsibility of the municipal authority, but the SEPA is in charge of the Environmental Code and will be able to steer applicants in the right direction.

Norwegian Pollution Control Authority
www.sft.no
Part of the Ministry of the Environment; responsible for supervision of installations with regard to protection of the environment. Contact for more regulatory and approvals information.

9 Support measures for bioenergy projects

New technological developments usually require a set of supporting measures to enable them to emerge from the research and prototype stages to become competitive products for sale in larger quantities on the open market. In most cases these measures are set up by government bodies in order to achieve certain goals that these technologies may support. Renewable energy systems (RES) are one option to reduce greenhouse gas emissions, and thus support the efforts of most countries' governments to decrease the speed of climate change. This chapter introduces the various support measures for bioenergy projects and provides detailed contacts and links where further information on national support measures can be found.

9.1 Introduction

A wide variety of measures to promote renewable energy systems are currently in use in various countries. In general, a successful renewable energy policy in each country does not depend on one single support mechanism but rather on a combination of various balancing effects.

These effects can be categorised into the following classes:

- political;
- legislative;
- fiscal;
- financial;
- administrative.

In addition, specific technology development programmes and educational efforts play a complementary role.

9.2 Support mechanisms for renewable energy systems

9.2.1 Supporting policies

The basis for a well-developed national renewable energy market is the incorporation of long-term renewable energy targets in the general energy policy of a country. In countries with a comparably large autonomy of regions or federal states, such as Germany, Austria or Spain, particular regional energy policies contribute here even further to the national policy.

Example 1
For the last 10 years Finish energy policy has been supporting the promotion of renewable energy systems, and in particular the utilization of biomass as a renewable energy source. In 1994 a national biomass strategy was launched, setting a target of an increase of 25% (~61 PJ/yr) in the use of biomass by 2005 compared with 1992. Five years later, in 1999, a renewable energy action plan was set up that increased these targets: for biomass it was raised to 114.5 PJ/yr.

Example 2
In 2003 the United States set up a biomass programme to support bioenergy issues, with two major long-term strategic goals: to reduce dependence on foreign oil by developing liquid fuels, and to create a domestic biomass industry. This is to be achieved by removing the barriers to cost-effectiveness and environmental viability.

9.2.2 Legislative measures

Looking at legislative measures to support renewable power generation, the starting point for green electricity is preferred access to the electricity grid at reasonable prices. The European Commission's Renewable Electricity Directive has laid a foundation in this regard, requiring transparent and reasonable charges for grid access. However, several EU member states implemented this in their legislation years ago.

In comparing current legislative support instruments for green electricity, basically three can be distinguished:

- feed-in laws;
- quota systems;
- competitive tendering.

Feed-in laws offer fixed revenues for each kWh of green electricity. Many EU member countries have feed-in laws in place, with Germany, Denmark and Spain showing particular success with this type of legislative measure.

Example
In Germany the Renewable Energy Sources Act (EEG) came into power on 1 April 2000 to replace the previous electricity feed-in law, which started the success story for green electricity by coming into power in 1990. The new law regulates two major aspects: it gives preferred access to the grid for electricity from renewable energy sources, and it sets for 20 years feed-in revenues for electricity that is fed into the national electricity grid. Moreover, it defines which energy sources are viewed as renewable energy sources and which are not. The electricity revenues vary depending on the renewable energy source and the capacity of the renewable energy system. Electricity from biomass receives revenues from €84 to €99 per MWh (2004); the smaller the system capacity, the higher the revenue.

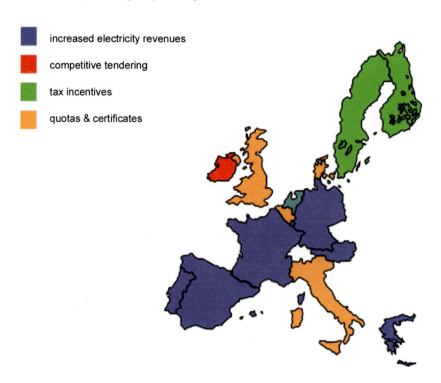

Figure 9.1.
Overview of major support mechanisms in the EU

A quota obligation system requires that the electricity supplier, producer, grid operator or consumer either generates or buys a certain share of green electricity.

Tradeable green certificates and penalties complement these quota obligations if the quota is not met. Currently in Belgium and the UK quotas are imposed on suppliers, in Italy they are imposed on producers, and in Sweden there are plans to introduce a consumer-based quota system.

Example
Since 1 April 2002 the Renewable Obligation (RO) has been in power in the United Kingdom. According to the RO, electricity suppliers are required to purchase a certain proportion of electricity from a range of renewable energy sources. During the first RO period, from April 2002 until March 2003, 3% of the supplied electricity has to originate from renewable energy sources. The proportion will increase to 10.4% by March 2011. If the supplier decides not to fulfil this requirement, a compensatory payment of £30/MWh has to be paid to the regulating body. The proof of purchase is done via Renewable Obligation Certificates (ROCs), which can also be traded among electricity suppliers.

The competitive tendering system commonly involves a call for tender for a given capacity of green electricity that offers a fixed price over a specified period of time for the winning bidder. Ireland (and formerly the UK) has a tendering system in place for electric capacity from wind and biomass. France had such system in place for wind energy, but has now switched to a feed-in system.

Both the quota obligation/certificate systems and competitive tendering are market-driven instruments, in contrast to the feed-in premium system.

9.2.3 Fiscal incentives

Fiscal measures include environmental taxes such as additional taxes on fossil fuels and CO_2 emissions, or tax exemptions for green electricity, but also tax incentives for investments in renewable energy systems (RES). Such instruments are intended to directly stimulate demand. Tax incentives are offered in a number of EU member states to complement the legislative measures. For example, in Germany and Sweden tax exemptions are offered for private investors investing in wind energy projects. In the Netherlands an accelerated depreciation scheme is offered to investors in RES to attract capital for new RES capacity. Only the Netherlands and the UK offer a tax redemption on the consumption or the generation of green electricity. France and Germany offer tax exemptions for the use of biofuels such as biodiesel in Germany and biodiesel and bioethanol in France.

Example
In the Netherlands every consumer of electricity has to pay a certain amount of 'ecological' tax called REB on each kWh consumed. The amount varies with the total annual consumption: large consumers (>10,000 MWh/yr) do not have to pay REB at all, whereas small consumers (<10 MWh/yr) are obliged to pay €63.9/MWh. If the consumer decides to purchase electricity from renewable energy sources, this REB is reduced by €29/MWh.

9.2.4 Subsidies, grants and loan programmes

Further financial support is given either by direct investment subsidies or by low-interest loans, partly with remittance of the remainder of the loan. Green certificates offer additional financial support for green electricity. Investment subsidies are being offered in most EU member states, but to different extents and focus. Owing to the maturity of the technology, in most countries wind energy is not eligible for investment subsidies anymore. However, this was the case for a number of years in many countries: for example, the Danish government supported wind projects and the development of this technology on a national level as part of their energy policy. Today, the strength of the Danish wind industry shows the result; it has become an important factor for the Danish economy. Photovoltaic (PV) systems have been and still are subsidized in most countries, as the technology is still rather expensive. However, in some countries with higher fixed premiums for electricity from PV systems, subsidies have been lowered or are available only for specific innovative installations, or programmes have ended. Throughout the EU, biomass systems are financially supported by direct investment programmes to support their further implementation. In addition to RES-related direct investment subsidies, subsidies have also come from EU structural funds, which offer subsidies to improve the infrastructure of certain regions or countries within the EU. Such grants were available in countries such as Portugal, Spain and Ireland, but also in regions in Austria and elsewhere.

Example 1
The Renewable Remote Power Generation Program in Australia offers grants of up to 50% of the capital costs of renewable energy installations that are operating off-grid.

Example 2
Low-interest loans are offered in the framework of the environmental programme of a German bank (Die Mittelstandsbank) for RES up to a debt capital proportion of up to 75% of the total capital costs. The interest on these loans is commonly about 2% lower than that of bank loans.

9.2.5 Administrative support for RES

Legislative, fiscal and financial measures are the most important factors in setting up a successful system of support mechanisms for green electricity. Nonetheless, once it gets to the actual implementation of RES a number of administrative hurdles need to be tackled: the two most important are approvals procedures and emissions standards (greenhouse gases, particles, noise etc.).

Example 1
In the Netherlands it is very hard to obtain approval for the installation of biogas systems with cogeneration engines. Although the Netherlands has a large capacity of manure from animal feedstock, the regulations for distributing the digested manure as a fertilizer are very strict.

Example 2
A positive example of a regulation that facilitates the implementation of RES is the demand placed on local authorities in Germany to designate certain areas in the municipal land-use plan that can be used for wind energy converters. This gives important planning guidance on the location of wind energy projects.

9.2.6 Technology development support

Promotion of the renewable energy market and sustainable implementation of renewable energies in the national economy are also promoted through technology development. Here, research and development subsidies and national research programmes are the most important and most-used instruments. Support at all stages, from research through demonstration to implementation, is necessary to obtain the required know-how and the qualified people. Another approach is to strengthen the national industry involved in renewable energy products and projects, thus supporting national economies by creating jobs and export potential. Excellent examples are the Danish and the German wind industry, but also worthy of mention are the photovoltaic industry in Germany and the Netherlands, and the competence in biomass in Austria, Finland and Sweden.

Example
The framework of the Renewable Energy Industry Development Program (REID) has been set up by the Australian government to support the Australian renewable energy industry by providing a competitive grants programme to companies that can

demonstrate that their projects will assist the development of that industry. REID provides A$ 6 million and has funded two previous rounds valued at over A$2 million.

9.2.7 Education and information

Last but not least education and information also contribute to the entire framework of a successful RES policy. In most EU member countries the establishment of national energy agencies has taken place, which fulfil functions such as offering information but also assisting in projects, thus actively implementing the energy policy. In some countries local energy agencies have also been formed that focus on specific local or regional interests.

Example
The European Soltherm Initiative is a central action network that was set up to stimulate the market growth of solar thermal products. It should lead to a major contribution to the EU's Campaign for Take-off goals by realizing 15 million m^2 of solar thermal collector area by 2004, thus supporting the EU's Kyoto targets for CO_2 emission reduction. In addition the initiative has been set up to create information exchange and education structures and an EU-wide network for the exchange of experience and know-how in the area of solar thermal applications.

9.3 General information on financial support

A sound financing structure is key to the economic success of a bioenergy project, and subsidies play an important role in this. However, the acquisition of funding from third parties such as public institutions or electric utilities is an art in itself, as each project has its own characteristics, and subsidy programmes are often related to only some of them. System size, the type of biomass used, the amount of heat produced by cogeneration, location – these are just a few of the parameters that influence selection of the appropriate programme. In general it is advisable to employ the help of an expert consultant who knows the way through the large number of support programmes and who has up-to-date knowledge on the actual status of such programmes.

The first step in acquiring third-party funding for bioenergy projects is to find out what kinds of institution provide support. These institutions can typically be classified into five groups:

- ministries and related institutions:
    - ministry for economic affairs;
    - ministry of agriculture and/or forestry;
    - ministry for environment;
    - ministry for research and development (for innovative projects);
- regional institutions:
    - ministries or institutions of federal states or regions;
- municipal institutions;
- independent organizations, e.g. environmentally focused foundations;
- energy utilities.

Despite the large number of support programmes, most of them have common aspects. In the following the most important aspects of applications to such support programmes are explained.

9.3.1 Project eligibility

This defines the type of system that is being supported by the particular programme, and the purpose that the system needs to fulfil. Generally a differentiation is made between biogas systems, combustion units fired by solid biofuels, and systems using liquid biofuels. Further to that, a difference is often drawn between systems that produce electricity only or heat only, and those that apply cogeneration. Constraints on system size or the eligibility of certain system components for financial support can also be specified here. For example, property costs are often excluded. Regional programmes clearly define the locations where bioenergy systems may be erected and operated.

9.3.2 Applicant eligibility

The group of bodies that are eligible for financial support varies depending on the type of bioenergy project and the support programme. Commonly a distinction is made between private people, enterprises of different sizes (SMEs etc.), agricultural or forestry enterprises, public institutions (e.g. universities) and power supply industries. In general, support programmes are designed for particular groups. Often programmes impose constraints on the eligibility of public bodies for funding, as the financial means are provided from public funds, and they are not supposed to flow back but to stimulate additional investments from non-public entities.

9.3.3 Essential qualifying (compliance) criteria

For an application to a funding programme for financial support to be successful, it is often necessary to comply with particular technical standards, or to apply specific project management rules. Commonly for bioenergy systems the following aspects may be covered:

- compliance with specified emissions limits;
- a requisite minimum efficiency or minimum heat utilization ratio;
- the type of permitted biomass.

It is also important to know that a project will not start before a particular grant has been applied for, or sometimes not until the grant has actually been received. However, in general planning can start before this.

9.3.4 Application form

In addition to the necessary application forms a number of annexes are commonly required to complement the information on the bioenergy project given in the form. Documents that enable the subsidizing institutions to evaluate the project proposal include:

- environmental impact assessment;
- information on socio-economic effects;
- contracts (e.g. for the supply of biomass, or for the rent of the property);
- bank approval of the project financing;
- land register extract;
- permits;
- cost breakdown;
- suppliers' tenders.

9.3.5 Type and level of funding

The financial support of bioenergy systems works either by grants or by low-interest loans. Tax incentives often complement these instruments.

The level of a grant depends on the eligible costs and the subsidy rate. A subsidy rate of 30% on the eligible investment costs means that the investor in a particular bioenergy project needs to finance 30% less of these costs. The subsidy rate varies depending on the size of the system, the type of biomass, the efficiencies and so on. Commonly subsidy rates range between 25 and 50%.

Example
A biogas system is supported by a low-interest loan of 4% p.a. including a partial debt remittance of 15% up to a certain power capacity of the cogeneration engine. For larger CHP engines the debt remittance is omitted, so that the support is granted only by the low-interest loan. This is a typical way of adapting the level of support to the economics of scale.

Often special support measures are offered if an old fossil-fuel-fired heating system is replaced by a biomass heating system – for example a wood pellet oven or centralized heating systems for a building or in combination with a solar thermal system – or if additional heat conservation measures such as improved insulation of the building are being applied.

9.3.6 Cumulation
Usually financing support from different state, regional or municipal support programmes can be combined in order to increase the level of support on a project. However, most support programmes limit the total subsidy rate on a project. A total subsidy rate of 50% of the total system costs is a common limit that can be found in most subsidy programmes.

9.3.7 Actual conditions for support programmes
Because support programmes are constantly changing, new ones are being set up and existing ones are being abolished, it was decided not to give any details about specific support programmes in this book. Instead we have provided a detailed list of sources of information where overviews of support programmes and their actual conditions can be obtained.

The following section provides information on support schemes in the field of bioenergy for different countries.

9.4 Further information on support measures in various countries

Because of the nature of policies and their implementation, the types and conditions of support measures and programmes for bioenergy systems are frequently changing. Therefore this section focuses on providing further sources of information rather than giving details of support measures.

A list with links to governmental and non-governmental organizations that provide information on policy regarding the use of biomass as a renewable energy carrier and their implementation and support measures is made available for several English-speaking countries – the UK, the USA, Canada, Australia, the Scandinavian countries, and the EU. These links will also help in acquiring information on regional programmes.

9.4.1 Sources of information in the UK
Links to governmental institutions, support programmes, associations and other sources of information.

9.4.1.1 GOVERNMENT
Department of Trade and Industry (DTI)
www.dti.gov.uk
Latest information on the DTI-funded programmes for renewable energy (calls for proposals etc.); also provides detailed information on policy regarding bioenergy.

Department for the Environment, Food and Rural Affairs (DEFRA)
www.defra.gov.uk
Latest information on support for energy crops.

Energy Savings Trust (EST)
www.est.org.uk
Information on energy efficiency and climate change; programmes for local authorities, consulting to small businesses.

UK Government Non-Food Use of Crops Research Database
cbaforms.maff.gov.uk/aims
Lists all government-funded R&D projects on non-food crops.

Office of Gas and Electricity Markets (OFGEM)
www.ofgem.gov.uk
Regulating organization for the gas and the electricity market; administers the Renewable Obligation (RO). Information on practical RO issues

The Carbon Trust
www.thecarbontrust.co.uk
The Carbon Trust is developing and implementing programmes to support low-carbon-emitting technologies

Enhanced Capital Allowance Scheme (ECA)
www.eca.gov.uk
Website set up by the Carbon Trust in collaboration with DEFRA and the Inland Revenue to provide information about the ECA Scheme.

Parliamentary Renewable and Sustainable Energy Group (PRASEG)
www.praseg.org.uk/
News on the practical implementation of support measures such as the Renewable Obligation.

9.4.1.2 FUNDING

DTI Support Programme
www.dti.gov.uk/renewable/geninfo.html
Information on UK government programmes supporting RES.

New Opportunities Fund
www.nof.org.uk
The New Opportunities Fund distributes National Lottery money, making available £50 million for offshore wind, energy crops and small-scale biomass heat projects.

Capital Grants Scheme (DTI)
www.dti.gov.uk/energy/renewables/support/capital_grants.shtml
Funding for demonstration projects in offshore wind projects, projects generating electricity from energy crops, and small-scale biomass heating schemes.

Landfill Tax Credit Scheme
www.entrust.org.uk
Purpose: to support community environmental projects and to encourage partnerships between landfill operators and local communities. ENTRUST, the Environmental Trust Scheme Regulatory Body, manages this scheme.

Landfill Tax Credit Scheme Bank Account: from The Co-operative Bank
www.co-operativebank.co.uk
Created a special account for organizations registered with ENTRUST.

DEFRA England Rural Development Programme
www.defra.gov.uk/erdp/erdpfrm.htm
Information on DEFRA's Rural Development Programme, including establishment grants for energy crops (Energy Crop Scheme – ECS).

Clear Skies Renewable Energy Grants
www.clear-skies.org
Information on the DTI's grant scheme for renewable energy systems.

Scottish Community and Householder Renewables Initiative (SCHRI)
www.est.org.uk/schri/
SCHRI offers grants, advice and project support to develop and manage new renewables schemes.

Community Renewables Initiatives (CRI)
www.countryside.gov.uk/communityrenewables/
Providing advice on how to set up RES projects, financing and funding, technology etc.

9.4.1.3 ASSOCIATIONS AND ORGANIZATIONS
British Bio Gen
www.britishbiogen.co.uk
Trade association for the British bioenergy industry; general information on bioenergy.

Renewable Power Association
www.r-p-a.org.uk
Trade association representing producers of renewable energy in the UK; broad information on renewable energy, policy and support measures.

Environmental Services Association
www.esauk.org
Trade association for companies providing waste management and associated environmental services.

National Assembly Sustainable Energy Group
www.naseg.org
Organization to promote sustainable development and renewable energy in Wales; special information on support programmes in Wales.
Western Regional Energy Agency and Network
www.wrean.co.uk
Energy agency in Northern Ireland promoting RES.

British Association for Biofuels and Oils (BABFO)
http://www.biodiesel.co.uk/
Organization dedicated to the promotion of transport fuels and oils from renewable sources.

9.4.1.4 R&D AND OTHER SOURCES OF INFORMATION
Biomass Pyrolysis Network (PyNe)
www.pyne.co.uk
A global network of active researchers and developers of fast pyrolysis of biomass; technology information.

Biomass Gasification Network (GasNet)
www.gasnet.uk.net
A global network of active researchers, developers and implementers of biomass gasification; technology information.

9.4.2 Sources of information in the USA
Links to governmental institutions, support programmes, associations and other sources of information.

9.4.2.1 GOVERNMENT
Department of Energy (DoE)
www.doe.gov
Energy policy, support programmes, links to further sources of information.

EERE – Bioenergy
www.eere.energy.gov/RE/biopower
Bioenergy part of the US DoE website on energy efficiency and renewable energy.

Biofuels Information Network
http://bioenergy.ornl.gov
Extensive information on bioenergy; hosted by the Oakridge National Laboratory.

9.4.2.2 FUNDING
www.science.doe.gov/grants/
Grants available through the US Department of Energy

Regional Biomass Energy Program (RBEP)
www.ott.doe.gov/rbep/
Information about five regional bioenergy programs administered by the Fuels Development Office within the DoE's Office of Transportation Technology. The programme has close affiliations with DoE's Office of Power Technologies.

Southeastern Regional Biomass Energy Program (SERBEP)
www.serbep.org
South-eastern part of the RBEP.

Great Lakes Regional Biomass Energy Program (GLRBEP)
www.cglg.org/1projects/biomass/index_frame.html
Great Lakes part of the RBEP.
Northeast Regional Biomass Program (NRBP)
www.nrbp.org
North-eastern part of the RBEP.

Pacific Regional Biomass Program (PRBP)
www.pacificbiomass.org
Pacific part of the RBEP.

Western Regional Biomass Energy Program (WRBEP)
www.westbioenergy.org
Western part of the RBEP.

US Department of Energy National Biofuels Program
www.biofuels.doe.gov
Programme dedicated to support of the development and application of biofuels.

The Energy Foundation
www.energyfoundation.org
Independent foundation supported by several other foundations to foster energy efficiency and clean energy; several support programmes for RES.

Database on State Incentives for RES
www.dsireusa.org
Project managed by the Interstate Renewable Energy Council (IREC), funded by the DOE; information on support measures in the different states of the US.

9.4.2.3 ASSOCIATIONS AND ORGANIZATIONS
Renewable Energy Policy Project
solstice.crest.org/bioenergy
Information, insightful policy analysis and news on RES; funded by DoE, EPA and several foundations.

American Bioenergy Association
www.biomass.org
A branch organization in the US for the biomass energy industry, advocating expanded production of power transportation fuels and chemicals from biomass.

Biomass Research and Development Initiative
www.bioproducts-bioenergy.gov
Coordinates R&D efforts in the field of biomass; managed by the National Biomass Coordination Office (DoE and Department of Agriculture – DA)

Biomass Energy Research Association
www.bera1.org
Information on bioenergy.

American Coalition for Ethanol
www.ethanol.org
Ethanol industry association; extensive information on ethanol including prices.

Renewable Fuels Association (RFA)
www.ethanolrfa.org
National trade association for the US ethanol industry; facts on policy, ethanol, plants and more.

National Biodiesel Board
www.biodiesel.org
National trade association representing the biodiesel industry; information on biodiesel, sources of biodiesel etc.
The Climate Ark
www.climateark.org
Portal on climate change and renewable energy; extensive information and links.

Green-e Renewable Electricity Certification Program
www.green-e.org
Certification of renewable electricity products. Providing information to consumers on green electricity. Administered by the non-profit Center for Resource Solutions.

Sustainable Energy Coalition
www.sustainableencrgy.org
Head organization of more than 30 associations active in the field of RES; provides news on RES.

9.4.2.4 R&D AND OTHER SOURCES OF INFORMATION
Alternative Fuels Data Center
www.afdc.nrel.gov
NREL data centre providing information on alternative fuels, listings of available alternative fuel vehicles, including an interactive fuel station mapping system.

National Renewable Energy Laboratories
www.nrel.gov
Information on RES (also bioenergy).

9.4.3 Sources of information in Canada
Links to governmental institutions, support programmes, associations and other sources of information.

9.4.3.1 GOVERNMENT
CANMET
www.nrcan.gc.ca
Performs and sponsors energy research, technology development and demonstration within Natural Resources Canada, a department within the Canadian federal government; information on funding.

NRCAN
www.nrcan.gc.ca
Federal government department specializing in sustainable development and use of natural resources.

Canadian Renewable Energy Network (CanREN)
www.canren.gc.ca
Created through the efforts of Natural Resources Canada (NRCan); information on all renewable energy technologies.

Office of Energy Efficiency (OEE)
www.oee.nrcan.gc.ca
Centre of excellence for energy efficiency and alternative fuels information.

9.4.3.2 FUNDING
National Biomass Ethanol Program (NBEP)
www.fcc-sca.ca
Administered by Farm Credit Canada (FCC) on behalf of Agriculture and Agri-Food Canada (AAFC).
Renewable Energy Deployment Initiative (REDI)
www.nrcan.gc.ca
Support programme for RES, particularly for highly efficient and low-emitting biomass combustion systems.

9.4.3.3 ASSOCIATIONS AND ORGANIZATIONS
Canadian Renewable Fuels Association
www.greenfuels.org
Information on ethanol, biodiesel etc.

9.4.4 Sources of information in Australia
Links to governmental institutions, support programmes, associations and other sources of information.

9.4.4.1 GOVERNMENT
The Australian Greenhouse Office
www.greenhouse.gov.au
Information on climate change, sustainable development and RES: links, funding programmes, comprehensive information etc.

Sustainable Energy Development Authority NSW
www.seda.nsw.gov.au
Information on RES in NSW: policy, support programmes, background information.

Sustainable Environment Authority of Victoria
www.seav.vic.gov.au
Information on RES in Victoria: policy, support programmes, background information.

Greenhouse Office of Victoria
www.greenhouse.vic.gov.au
Information on the Victorian Greenhouse Gases Reduction strategy.

Energy SA Sustainable and Renewable Energy
www.sustainable.energy.sa.gov.au/home/home.htm
Information on RES in Southern Australia: policy, support programmes, background information.

Western Australian Government: Office of Energy
www.energy.wa.gov.au
Information on RES in Western Australia: policy, support programmes, background information.

Department of Infrastructure, Energy and Resources, State Government of Tasmania
www.dier.tas.gov.au
Information on RES in Tasmania: policy, support programmes, background information.

Department of Business, Industry and Resource Development of Northern Territory
www.dme.nt.gov.au
Information on RES in Northern Territory: policy, support programmes, background information.

Sustainable Energy Development Office of Western Australia
www.sedo.energy.wa.gov.au
Information on RES in Western Australia: policy, support programmes, background information

9.4.4.2 FUNDING
Renewable Energy Rebate Program
www.dme.nt.gov.au
Support for Remote Area Power Supply (RAPS) systems incorporating renewable energy.

Government Programs for RES
www.greenhouse.gov.au/renewable/government.html
Information on a number of national government programmes to support RES.

Western Australia RES Programs
www.sedo.energy.wa.gov.au
Information on a number of support programmes directed to different RES: R&D and market penetration.

9.4.4.3 ASSOCIATIONS AND ORGANIZATIONS
www.users.bigpond.net.au/bioenergyaustralia/Home.htm
A government-industry forum to foster and facilitate the development of biomass for energy, liquid fuels, and other value added bio-based products.

Biodiesel Association of Australia
www.biodiesel.org.au
Promotion of biodiesel in Australia: background information, links.

Australian Biofuels Association
www.australianbiofuelsassociation.org.au
Support and lobby for Australian growers and processors of biomass feedstocks, domestic biofuels producers, biofuel distributors, research and development organizations; comprehensive information on biofuels.

9.4.5 Sources of information in Scandinavia
Links to governmental institutions, support programmes, associations and other sources of information.

9.4.5.1 GOVERNMENT
National Danish Energy Information Centre
www.energioplysningen.dk
More information on RES, background, policy and links in Denmark.

Danish Energy Authority
www.ens.dk
Information on energy policy in Denmark, including legislation (e.g. emissions standards), information on RES, support programmes, links in Denmark.

Finnish Ministry of Trade and Industry
www.ktm.fi
Information on RES, support programmes, links in Finland.

National Technology Agency of Finland
www.tekes.fi
Information on RES, support programmes, links in Finland.

Norwegian Energy Agency
www.enova.no
Part of Royal Norwegian Ministry of Petroleum and Energy. Enova's main mission is to contribute to environmentally sound and rational use and production of energy; information, support programmes and other financial instruments and incentives.

Swedish Energy Agency
www.stem.se
Information on RES, support programmes, links in Sweden.

9.4.5.2 ASSOCIATIONS AND ORGANIZATIONS
Danish Biomass Association (DANBIO)
www.biomass.dk
Information on bioenergy and support programmes in Denmark.

Svenska Bioenergiföreningen (SVEBIO)
www.svebio.se
The Swedish biomass association: organizer of the World Bioenergy Conference 2004. Information on all bioenergy-related topics in Sweden.

Norwegian Bioenergy association (NoBio)
www.nobio.no
Information on bioenergy in Norway; includes list of bioenergy fuel suppliers.

Finnish Bioenergy Association (FINBIO)
www.finbioenergy.fi
Comprehensive information on bioenergy, companies, technology, statistics, links and more.

Swedish Association of Pellet Producers (PiR)
www.pelletsindustrin.org
Members own and operate 16 out of more than 20 production plants in the country. Information on pellets and producers.

Danish Centre for Biomass Technologies
www.videncenter.dk
Detailed information on bioenergy in general, technology; particularly interesting for straw.

9.4.5.3 R&D AND OTHER SOURCES OF INFORMATION
Nordic Energy Research
www.nefp.info
Focus: integration of the energy market, renewable energy sources, energy efficiency, the hydrogen society and consequences of climatic change on the energy sphere.

Technical Research Centre of Finland
www.vtt.fi
Comprehensive information on bioenergy.

Teknologisk Institut
www.teknologisk.dk
Test centre for small-scale biomass combustion units; includes lists of approved boilers.

9.4.6 Sources of information in other English-speaking countries
Links to governmental institutions, support programmes, associations and other sources of information.

9.4.6.1 SOUTH AFRICA
Department of Minerals and Energy
www.dme.gov.za
Information on RES policy and support programmes in South Africa.

9.4.6.2 NEW ZEALAND
Energy Efficiency and Conservation Authority (EECA)
www.eeca.govt.nz
Information on RES policy and more in New Zealand.

Bioenergy Association of New Zealand (BANZ)
www.bioenergy.org.nz
Background information on bioenergy, technology, publications etc.

9.4.7 Sources of information at EU level
A Global Overview of Renewable Energy Sources (AGORES)
www.agores.org
EC web site providing information on sources of European funds, RES policies, general technology information incl. project descriptions, national key players in the EU member states and publications on RES.

CORDIS
www.cordis.lu
Information server of the EC, providing information on R&D, innovations, support programmes and more.

European Biomass Association (AEBIOM)
www.ecop.ucl.ac.be/aebiom
Head association of national biomass associations; political organization to strengthen the European bioenergy market development; general information, links, position papers, newsletter.

European Renewable Energy Centres Agency (EUREC)
www.eurec.be
Economic interest grouping to strengthen and rationalise European R&D efforts in renewable energy technologies.

9.4.8 Other sources of information on bioenergy

9.4.8.1 BIOMASS DATA
Phyllis
www.ecn.nl/phyllis/
Dutch database of biomass and waste; analysis data on a large number of different biomass streams.

BIOBIB
www.vt.tuwien.ac.at/biobib/search.html
Austrian database of biofuels; managed by the Technical University of Vienna.
Climate Neutral Gaseous and Liquid Energy Carriers (GAVE)
www.gave.novem.nl
Dutch government programme; general and detailed information on biofuels.

9.4.8.2 INTERNATIONAL
International Energy Agency (IEA)
www.iea.org
Information on all energy-related issues.

IEA Bioenergy
www.ieabioenergy.com
Bioenergy section of the IEA: detailed information on bioenergy R&D and state-of-the-art technology; many reports to download.

Renewables Information Database
www.iea.org/statist/renew.htm
Provides statistics on renewable energy in OECD member states.

Centre for Analysis and Dissemination of Demonstrated Energy Technologies (CADDET)
www.caddet-re.org
International information network to provide managers, engineers, architects and researchers with information about renewable energy and energy-saving technologies.

AFB NET
www.vtt.fi/virtual/afbnet/
European bioenergy network; provides detailed information on technology, potentials, national activities on bioenergy.

European Energy Crops InterNetwork
www.eeci.net
Network supports and bundles and disseminates information on research, development and implementation activities of energy crops.

Index

Page references in italics refer to figures and tables

agriculture 1, 3, 12, 19, 29, 215, 217
air humidity 25, 170, 183, 184
alcohols 18, 22, 104, 108, 109, 114
Amergas gasifier, Netherlands 219, 220
ammonia (NH_3) 51, 58, 59, 75
anaerobic digesters
  advantages 61–2
  batch processes 63–4, 66
  centralized (CADs) 54
  in CHP systems 55, 72–3, 76, 80, 82–6, 90–1, 95, 97
  components 54, 55, 66–77, 78, 93–6
    continuous processes 63, 64, 66
  costs 65, 67, 82, 93–9, 99–100
  digester heating 54, 67, 73
  digester insulation 54, 67, 68, 85, 95
  digester tanks 54, 63, 66, 71, 92, 94, 95
    digesters 54, 55, 63–4, 76, 82–3
  dimensioning 74, 82–7
  farm scale 54, 55, 56, 64, 71
  gas engines 54, 55, 63, 71, 72–3, 76
  heat exchangers 67, 72, 95, 96
  heat output 79, 81, 92, 99
  heating arms 64
  horizontal digesters 64, 65, 66, 67
  industrial scale 56, 64
  maintenance 92, 93, 98
  malfunctions 92
  measurement and control equipment 74–6
  meters 74
  mixing 55, 58, 64, 65, 69–70, 87, 96
  operation 90, 91–2, 98
  organic load 57, 60
  piping 54, 67, 68–9, 70, 76, 85–6, 95, 96
  post-digestion tank 63, 74, 83, 94
  pre-treatment 54–5
  processes 54, 63–4, 66
  pumps 54, 68, 69, 70, 86–7, 96
  retention times 57, 58, 65
  safety issues 76, 77, 90, 96
  sanitation requirements 55, 63, 64, 85–6
  sanitation tanks 67
  screw propellor system 69, 70
  semi-continuous process 64, 66
  sizes 53, 54, 56
  small scale 53, 56, 64
  start-up 89–91
  stirring devices 54, 69–70
  storage systems 54, 55, 66, 70–2, 74, 83, 96
  substrate level indicators 74
  system layouts 54–5, 63–6
  temperature 57, 58, 67, 74
  upright digesters 64, 65, 67
  ventilation 92
  see also co-substrates; substrates
anaerobic digestion (AD)
  bacterial activity 57–8, 67
  biogas production 54, 55, 59, 60, 61, 71–4, 75, 89
  biological process 57–8
  economic considerations 53, 65, 67, 79, 82, 89, 93–100
  electricity generation 53, 54, 73, 76, 79, 80, 90, 99
  of manure 53–63
  projects
    checklists 88
    construction and engineering 80, 87, 88–9, 95, 97
    cost-benefit analysis 78, 99–100
    creation 77, 78–80
    economic considerations 78–9, 82, 89, 99–100
    feasibility studies 77, 78, 80–7, 97
    funding 79, 88, 97–8
    legal considerations 61, 62, 79, 80, 87–8, 90, 234, 235
    permits 61, 79, 80, 87, 88, 90, 234
    planning 53, 77–89
    preparation 77, 78, 87–8
    realization 77, 78, 88–9
    revenue 99–100
    start-up 53, 77, 78, 88–91
    suppliers 53, 80, 81, 87, 90
    support programmes 97–8
Australia 239–40, 246, 254–5
Austria 5, 6, 102, 243, 246

bacteria
  in anaerobic digestion 57–8, 67
  in biogas production 15, 23, 54, 75, 89
  in desulphurization 73
  ideal living conditions 57–8, 75
  in manure 62, 89
Belgium 5–6, 102, 245
biodiesel
  availability 34
  in blends 106, 108, 114
  in CHP systems 114, 224
  in conventional engines 3, 28, 108–9, 115–16, 224
  from cooking fats 106, 108
  corrosive properties 108–9, 115
  economic considerations 113, 114
  energy efficiency 105
  environmental issues 18, 114
  heating oil equivalent 40
  in heating systems 40
  market development 114
  in mobile applications 3, 18, 28, 104, 108–9, 115–16
  performance issues 23
  production 106
  quality standards 27–8

in stationary applications 114, 224
technical properties 108–9
from vegetable oils 22, 104, 108
viscosity 40, 108
bioenergy products 29–34
bioenergy projects
applicant eligibility 248
boundary conditions 1, 5–6, 231–42
compliance criteria 248
environmental issues 231
financial support
competitive tendering 245
cumulation 249
grants 211, 245, 246, 247, 248, 250, 252
information sources 247–58
loans 6, 79, 88, 97, 211, 245, 246, 248
quotas 5–6, 245
subsidies 5–6, 99–100, 103, 226, 228, 245–6, 247, 248, 249
tax incentives 6, 103, 245, 248
legal considerations 231–42, 244–5, 246
permits 231, 232–6, 236–42
project eligibility 247–8
safety issues 233
support programmes 6, 112, 114, 233, 243–58
technical rules 76, 232, 233, 235
see also under anaerobic digestion; biofuels; CHP systems; wood energy projects
bioenergy sources
carbon dioxide (CO_2) neutral 2, 12, 23, 217
in environmental protection 1–2, 3, 4, 15, 23, 231, 235
forms of 15–17
gaseous 1, 16, 18, 23–4, 28–9 see also biogas; synthesis gases; wood gas
job creation 23, 104
liquid 1, 16, 18, 22–3, 27–8, 51, 101–18 see also biofuels
for mechanical energy 17, 18, 47
solid 1, 16, 17, 19–22, 24–7 see also stem products; straw; wood
utilization 16–19, 29, 42, 201, 217, 218
biofuels
advantages 103–4
applications 104–6
in CHP systems 45–8, 117–18
costs 103, 106
economic considerations 103, 104, 112–13, 114
energy efficiency 104, 105, 110, 112, 114
environmental issues 18, 101, 103, 112, 114, 115, 116–17
incentives for use 103
market development 113–15
mobile applications 115–16
production 106–12
projects 101, 110, 116
from recycled materials 107
research 16, 51, 104, 109, 110, 111, 112, 117, 118
stationary applications 3, 101, 114, 115, 116–17, 234 see also under CHP systems
in transport sector 3, 35, 51, 101–6
see also biodiesel; ethanol; hydrogen; methanol; synthesis gases; vegetable oils
biogas
from anaerobic digestion 54, 55, 59, 60, 61, 71–4, 75, 89
availability 34
biogas plants see anaerobic digesters

from biowaste 15, 28, 56, 234
calorific value 28, 84
in CHP systems 46, 53, 54, 55
in combustion engines 46, 71, 72–3
composition 59, 75
as cooking fuel 55
corrosive properties 29, 41, 51, 72, 75
desulphurization 73–4, 75
flame propagation 41
from landfill sites 18, 19, 23
measuring 74–5
in natural gas grids 51–2, 55
pressure lines 51–52
from recycled residues 19
safety issues 76, 77, 90
stationary applications 34
storage 55, 66, 71–2, 83–4, 92, 94
as vehicle fuel 51, 55
yields 59, 60, 61, 71, 83, 90
biogenic gases see biogas; wood gas
biomass
aggregate states 16, 17, 19, 35
calorific value 24, 25, 35
CO_2 stored 7, 8, 9, 12–14
as an energy carrier 1–4, 5, 7, 218, 232, 234
legal considerations 232, 233, 234
potential 4, 5, 210
types of 1, 4, 14
see also manure; plants; wood
biowaste 15–16, 28, 56, 111, 228, 234
boiler rooms
for log-fired central heating boilers 152, 154, 155, 156
soundproofing issues 178
for wood pellet boilers 158, 164, 183
for woodchip boilers 167
boilers see gas boilers; oil-fired boilers; solid fuel boilers
Brazil 34, 114, 116
Brigon CO_2-indicators 75

Canada 238–9, 253–4
carbon cycle 8, 11–12
carbon dioxide (CO_2)
in biogas 23, 28, 51, 59
in biomass 7, 8, 9, 12–14
in climate change 1–2, 8–11
in combustion process 12, 36
emissions 2, 11, 13, 101–2, 103, 104, 105
from gasification 24, 29, 222, 223
in photosynthesis 7, 8, 9, 12
carbon monoxide (CO)
in combustion process 36, 37, 38, 39, 234
as a flue gas 136, 143, 234
from fossil fuels 73, 114
from gasification 24, 29, 110, 111, 112, 218, 223
central heating cookers 142–5
central heating systems
central heating boilers 39, 127, 152–69
central heating cookers 142–5
log-fired boilers 152–58
in residential buildings 121–2, 128, 130, 142–5, 152–69
from solid fuel boilers 39, 127, 152–69
wood pellet boilers 158–66, 169
woodchip boilers 166–8
central pellet boilers see wood pellet boilers
check valves 69, 76
chimneys

diameter 137, 156, 164, 179
dimensioning 131, 137, 156, 164, 178, 180
draught limiters 178
in fireplaces 131, 132, 134
flue gases 137, 144–5, 153, 178
flue pipes 134, 150, 156, 164, 180–1
insulation 137, 156, 164, 179, 180, 181
materials 179
moisture resistance 179
noise insulation 164, 178, 181
problems/solutions 132, 134, 138
in solid fuel boilers 156, 160, 164, 178–81
soot fire resistance 137, 144, 156, 164, 179
in stoves 134, 137, 138, 150
surface finish 179
temperature resistance 156, 164, 179
ventilation 156, 164
CHP systems
   with anaerobic digesters 55, 72–3, 76, 80, 82–6, 90–1, 95, 97
   with biodiesel 114, 224
   with biofuels 45–8, 115, 116, 117–18
   with biogas 46, 53, 54, 55
   with combustion engines 45–8, 49
   dimensioning 42–5, 84
   economic considerations 42, 95, 116, 117–18, 226
   for electricity generation 42, 45–51
   emissions 225, 234–5
   energy efficiency 42, 116, 117
   environmental issues 116–17
   with gaseous fuel 44–8 see also biogas; synthesis gases
   generators 42, 49
   heat exchangers 42, 45, 72, 95
   heat production 42, 43, 86
   hydraulic integration 44
   maintenance 88, 90, 92, 98, 116, 118
   malfunctions 92
   noise insulation 44
   operating times 43, 116
   power output 44
   projects 2, 117, 118
   with solid fuel 16, 44
   with steam 42, 47–9
   with storage tanks 44–5
   with synthesis gases 217, 224–5
   with turbines 42, 48–51
   vegetable oil use 116, 117, 118
climate protection and climate change 1–2, 8–11, 10, 42
(co-)digestates see digestates
co-substrates in anaerobic digesters
   availability 79
   biogas yield 60, 61
   chemical content 62
   co-digestion 57, 60, 79, 96
   contracting arrangements 81–2
   costs 96, 98–9
   energy crops 60, 61
   feeding systems 71
   legal considerations 61, 90, 234–5
   nutrients 79, 81, 82, 98
   particle size 58, 71
   pathogens and seeds 62–3
   pH value 57, 60, 75
   physical impurities 62
   pre-treatment 54–5
   silage 60, 61
   size reduction 71
   sludge 63
   slurry 60, 61, 62
   storage 54, 71
   vegetable waste 61
   see also digestate
cogeneration plants see CHP systems
combustion
   CO_2 released 12, 36
   cold combustion 12
   defined 217
   emissions 12, 27
   endothermic reactions 36, 37
   energy density 12
   exothermic reactions 36, 37
   external 45, 47
   of fuel/gas/air mixtures 18, 46, 49
   of gaseous fuels see gaseous fuel combustion
   internal 45
   of liquid fuels see liquid fuel combustion
   of solid fuels see solid fuel combustion
   temperature requirements 25, 27, 36, 38
   see also combustion engines; combustion systems
combustion engines
   air/gas/fuel mixtures 18, 44, 73
   with biodiesel 3, 27, 108–9, 115–16, 224
   with biogas 46, 71, 72–3
   in CHP systems 45–8, 49
   dual fuel engines 18, 44, 73
   as emergency power units 73
   ethanol used 109
   fluorocarbon rubber components 108–9, 115
   with fuel cells 42, 50–1, 72, 224, 225
   gas-diesel engines 45, 46, 224
   gas piston engines 72–3
   gas spark ignition engines 45, 46
   injection engines 45, 46, 47, 109, 224–5
   Stirling engines 42, 45, 47, 224, 225
   with synthesis gases 217, 224–5
   vegetable oils used 27, 115, 117
combustion systems
   combined heat and power systems see CHP systems
   emissions 38–9
   for heating see heat and heating systems
   for stem products 27
   storage tanks 39, 44–5
   for wood-burning 17, 21, 29–30, 31–2, 33
   see also types of system
corrosion
   biodiesel causes 108–9, 115
   biogas causes 29, 41, 51, 72, 75
   by methanol 111
   in pipes 76, 108
   in stoves 140
   in wood boilers 170, 181

decomposition 11, 12, 16, 36, 37, 57, 222
Denmark 240, 241, 244, 245, 246, 255, 256, 257
diesel 73, 90, 101, 103, 104, 105, 110, 112
digestate in anaerobic digesters
   chemical content 62, 75, 234
   as fertilizer 56, 62, 74, 79, 82, 99, 233, 235
   legal considerations 62, 79, 235
   pathogens and seeds 62, 63
   storage 55, 66, 71, 74, 83

economic considerations
   of anaerobic digestion systems 53, 65, 67, 78–9, 82, 89, 93–100

of biofuels 103, 104, 112–13, 114
of CHP systems 42, 95, 116, 117–18, 226
financial support
   competitive tendering 245
   cumulation 249
   grants 211, 245, 246, 247, 248, 250, 252
   for green electricity 5, 6, 78, 99, 228, 244–5
   information sources 247–58
   loans 6, 79, 88, 97, 211, 245, 246, 248
   quotas 5–6, 245
   subsidies 6, 99–100, 103, 226, 228, 245–6, 247, 248, 249
   tax incentives 5–6, 103, 245, 248
of fossil fuel use 102, 103, 112, 113
funding for projects 79, 88, 97–8, 214, 248, 249
of gasification 226–9
of usable manure 79
of wood energy projects 201, 210–11
education 6, 231, 247
electricity
   from anaerobic digestion systems 53, 54, 73, 76, 79, 80, 90, 99
   from CHP systems 42, 45–51
   constant nominal production 73
   demand driven 73
   economic considerations 5, 6, 78, 99, 228, 244–5
   feed-in 81, 231
   feed-in tariffs 5, 6, 79, 81, 228, 244, 245
   financial support 5, 6, 79, 99, 228, 244–5
   from gaseous fuels 42, 45–8, 217, 223, 224
   from gasification 217, 223, 224
   grid access permits 232, 236
   grid connection 76, 80, 81, 90, 213, 231, 232
   legal considerations 231, 244
   peak demands 71, 81
   safety issues 76
   in storerooms 164, 193, 194
   support programmes 234
   as vehicle fuel 104
emissions
   carbon dioxide 2, 11, 13, 101–2, 103, 104, 105
   chlorine 27
   from CHP systems 225, 234
   from combustion systems 38–9
   from fossil fuels 2, 10, 23, 73, 101, 114
   from gasification process 225
   of greenhouse gases 1–2, 13, 241–2, 245
   legal considerations 27, 80, 234, 235, 236, 248
   nitrogen 27
   from vehicles 101–2, 103, 104, 105, 110
energy crops
   in anaerobic digestion 60, 61
   conversion 16
   as an energy source 15
   examples 15
   Miscanthus sinensis 15, 26
energy efficiency
   of biofuels 104, 105, 110, 112, 114
   of CHP systems 42, 116, 117
   of gasification process 227–9
environmental issues
   of bioenergy sources 1–2, 3, 4, 15, 23, 231, 235
   of biofuel use 18, 101, 103, 112, 114, 115, 116–17
   environmental impact assessments 231, 235, 237
   environmental permits 79, 90, 237, 238, 239, 240, 241–2

   environmental taxes 245
   of fossil fuel use 23, 42, 101, 103, 116–17
   of gasification 217, 218, 225
   of heating systems 119, 125, 163, 172, 208, 209
   see also emissions; pollution
ETBE 106
ethanol
   in blends 18, 28, 34, 109, 114, 116
   in CHP systems 115
   distillation 109
   economic considerations 113
   energy efficiency 105
   environmental issues 114
   from lignocellulose materials 109
   market development 114–15
   in mobile applications 18, 28, 34, 104, 116
   production 22, 106, 109, 114
   pure 18, 28, 34, 109, 114, 116
   quality standards 28
   from sugar-based materials 106, 109, 114
   technical properties 109
Europe
   anaerobic digesters 56
   biodiesel use 18, 34
   bioenergy use 4–6, 243
   economic incentives 5–6, 112, 244, 245–6
   ethanol use 18, 34, 114, 116
   information sources 247
   vegetable oil production 107
   see also individual countries
European Soltherm Initiative 247
European Union (EU)
   bioenergy policy 2, 5, 244–6
   biofuel production and use 2, 102–3, 104
   feed-in tariffs for bioenergy 6
   information sources 257
   vehicle emissions 101–2, 104
   see also Europe

fatty acids 75, 104, 108 see also biodiesel
Finland 5, 6, 233, 240, 241, 243, 246, 256, 257
fireplaces
   air circulation 131, 132, 133
   ash pans 132, 134
   closed 132–4
   dimensioning 131
   fireboxes 132
   open 130–2
   problems/solutions 132, 134
   specifications 130, 133
   structural requirements 131, 133–4
   see also chimneys
Fischer-Tropsch synthesis 111
food industry 15, 103, 106, 107, 108
forests and forestry 1, 12, 13, 19, 29, 212, 215, 217
fossil fuels
   economic considerations 102, 103, 112, 113
   emissions 2, 10, 23, 73, 102, 114
   as an energy source 1–6, 8, 31, 35, 42, 90, 102
   environmental issues 23, 42, 101, 103, 116–17
   see also oil; petrol
France 5, 18, 102, 114, 245

gas see biogas; synthesis gases; wood gas
gas boilers
   atmospheric boilers 41
   in CHP systems 44–5
   condensing boilers 41, 44–5
   low-temperature 41, 44
gaseous fuel combustion

biogas 46, 71, 72–3
  for electricity generation 42, 45–8, 217, 223, 224
  flames 37, 41
  in gas boilers 41, 44–5
  synthesis gases 217, 218, 224–5
gasification
  applications 223
  by-products 225
  carbon dioxide produced 24, 29, 222, 223
  carbon monoxide produced 24, 29, 110, 111, 112, 218, 223
  co-current gasifiers 220, 221–3
  in combustion process 36, 38
  countercurrent gasifiers 219
  defined 217
  downdraught gasification 220
  drying of the fuel 222
  dust contamination 29, 110, 112, 221, 223
  economic considerations 226–9
  economic viability 227–9
  electricity generation 217, 223, 224
  emissions 225
  energy efficiency 227–9
  energy requirements 110, 229
  entrained-flow reactors 221
  environmental issues 217, 218, 225
  financial revenue 227
  financing 227
  fixed-bed gasifiers 219, 221, 222–3
  fluidized-bed gasifiers 219, 220, 221
  fuel 219, 221
  gasification technology 111, 218, 219–3, 227, 229
  gasifiers 218, 219–23
  heat generation 36, 38, 217, 223, 224, 227–8
  hydrogen produced 24, 29, 110, 112, 222, 223
  manure as fuel 28, 221
  methane produced 24, 29, 110, 222, 223
  methanol produced 110, 111
  nitrogen produced 223
  oxidation 222
  process 36, 38, 217, 218–9, 222–3
  pyrolytic decomposition 222
  research 109, 110, 111, 112, 217, 222
  support programmes 227
  synthesis gases 24, 29, 109–11, 217, 218, 222–25
  tar contamination 29, 36, 110, 218, 219, 220, 224, 225
  thermochemical process 23, 24, 28, 36, 51, 109, 110–11, 217–18
Germany
  anaerobic digestion projects 99
  bioenergy use 5, 213, 243
  biofuel use 102, 104, 111, 114, 115, 116, 245
  financial measures 245, 246
  gasification technology 111, 219, 227
  heat demand of buildings 120
  legal measures 234, 244, 246
greenhouse effect 1–2, 8–10

heat and heating systems 42–5
  auxiliary systems 151
  from bioenergy sources 1, 17, 35–52
  bivalent systems 126, 127–8
  central heating 95, 121–2, 128, 130, 142–5
  centralized systems 129, 130
  combined heat and power systems see CHP systems
  dimensioning 121–5, 176–7
  energy requirements 43, 110, 120, 121, 202
  environmental issues 119, 125, 163, 172, 208, 209
  expansion tanks 175–77
  from gasification 36, 38, 217, 223, 224, 227–8
  hearth appliances 122–5, 128, 130–51, 152
  heat demand 79, 81, 119–28, 129, 162, 171
  heat generators see fireplaces; solid fuel boilers; stoves
  heat output 40, 122, 124, 125, 128, 129
  heating networks 17, 35, 43, 204, 207, 210, 213
  hot water provision 119, 125, 139, 149, 158, 162, 171, 208
  large-scale 17, 201–15, 233, 235
  legal considerations 231–2, 233, 235–6
  malfunctions 92
  monovalent systems 126–7
  piping 67, 85–6
  radiators 169, 170, 171, 177
  residual heat 56, 85, 86, 99
  safety equipment 175
  seasonal demand 125–28
  small-scale 17, 119–200, 233–4, 233, 235–6
  from synthesis gases 217, 223, 224
  thermostats 39, 154, 170
  with wood 17, 20, 119, 124–5, 128–30, 201–15
heat exchangers
  in anaerobic digesters 67, 72, 95, 96
  in central heating cookers 142, 144
  in CHP systems 42, 45, 72, 95
  in log-fired central heating boilers 153
  in solid fuel boilers 153, 159, 162, 165
  in tiled stoves 146, 147
hydrogen (H_2)
  in biogas 59
  energy efficiency 112
  environmental issues 112
  as a fuel 110, 112
  from gasification 24, 29, 110, 112, 222, 223
  in vegetable oils 107
  in vehicle fuel 51–52
hydrogen sulphide (H_2S) 28, 51, 59, 73, 75

information sources 236, 237, 238, 247–60
insulation
  of anaerobic digesters 54, 67, 68, 85, 95
  calculations 120
  of chimneys 137, 156, 164, 179, 180, 181
  costs 95
  dimensioning 85
  of fireplaces 134
  mesh 67, 68
  mineral wool 67, 68
  noise insulation 44, 164, 178, 181
  for piping 69
  polystyrene 67, 68
  polyurethane foam 67, 68
Ireland 245
Italy 5–6, 102

Joos gasifier 221, 222

landfill sites 18, 19, 23
legal considerations
  of anaerobic digestion projects 61, 62, 79, 80, 87–8, 90, 234, 235
  of bioenergy projects 231–42, 244–5, 246
  biomass-related 232, 233, 234
  building regulations 119

of electricity feed-in 231, 244
for emissions 27, 80, 234, 235, 236, 248
permits
  anaerobic digestion projects 61, 79, 80, 87–8, 235
  applications 236
  bioenergy projects 231, 232–36, 236–42
  building permits 232, 236–7, 238, 239, 240, 240–1
  environmental 79, 90, 237, 238, 239, 240, 241–2
  grid access permits 232, 236, 237, 239–40, 240–1
  heating systems 231–2, 233, 235–6
of waste disposal 16, 233, 234, 237
of wood energy projects 201, 212–16
lignin 62, 109
lignocellulose materials 22, 106, 109
liquid biofuels see biofuels
liquid fuel combustion
  biodiesel 40
  carbonization results 40
  flames 37
  oil-fired boilers 40, 41
  vegetable oils 40, 41
log-fired central heating boilers
  air circulation 155–6
  ash pans 156
  chimneys 156
  fans 154, 156
  flue gases 153, 156
  flue pipes 156
  heat exchangers 153
  heat output 155
  maintenance 156
  operation 156–7
  problems/solutions 157–8
  servicing 157
  specifications 155
  structural requirements 155–6
  thermostats 154
logs 29
  ash content 26, 33
  calorific value 26
  in central heating cookers 142
  in combination boilers 168
  energy content 26
  energy density 33, 182
  in hearth appliances 124
  heat output 155
    log-fired central heating boilers 152–58
    measurements 25, 26, 33
    in solid fuel boilers 29, 33, 152–58, 181
  splitting 33, 182
  storage 33, 181–2
  water content 26, 33, 181

maintenance
  of anaerobic digesters 92, 93, 98
  of central heating cookers 145
  of CHP systems 88, 90, 92, 98, 116, 118
  of solid fuel boilers 156
  of stoves 138, 140–1
manure
  acid contents 58
  in anaerobic digestion see anaerobic digesters
  antibiotics 58
  biogas yield 59
  chicken manure 59, 221

composition 57–8, 59, 61–2, 79
cow manure 59, 60, 79
digested 61–2
disinfectants 58
dry matter content 59, 60, 79
economic considerations 79
as gasification fuel 28, 221
methane production 54, 57, 63
odour 62
organic matter content 59, 60
pathogens and seeds 62
pig manure 59, 60, 79
piping 54, 68–9, 96
pumps 86–7
storage 54, 64, 65, 70, 74
mechanical energy from bioenergy sources 17, 18, 47
methane (CH_4)
  in biogas 15, 23, 28, 51, 59, 73, 75, 90
  calorific value 28
  from fermentation 15, 23, 28, 63
  from gasification 24, 29, 110, 222, 223
  as a greenhouse gas 8, 9
  from manure 54, 57, 63
  in vehicle fuel 51
methanol
  in biodiesel 27, 108, 111
  corrosive properties 111
  economic considerations 113
  energy efficiency 105, 112
  environmental issues 112
  as a fuel 22, 104, 111–12
  from gasification 110, 111
  heating value 111
  production 22, 106
  pure 111
  risks 112
  technical properties 111–12
methyl esters 104 see also fatty acids
mineral oil-based fuel see diesel; oil; petrol
MTBE 106

The Netherlands 219, 221, 234, 245, 246
New Zealand 257
nitrogen (N_2) 29, 51, 58, 59, 61, 223
nitrogen oxides (NO_x) 39, 112
Norway 241, 242, 256

oil 113
oil-fired boilers 40, 41
organic by-products 15, 16
organic waste see biowaste
oxygen (O_2)
  in biogas 28, 51
  in carbon cycle 12
  in combustion process 36
  in photosynthesis 7, 8, 9
  in vegetable oils 107, 118
  in vehicle fuel 51

petrol 102, 104, 110, 112
photovoltaic (PV) systems 245–6
piping
  in anaerobic digesters 54, 67, 68–9, 70, 76, 85–6, 95, 96
  corrosion issues 76, 108
  dimensioning 85–6
  flue pipes 134, 150, 156, 164, 180–1
  for heating systems 67, 85–6
  insulation 69

materials 67, 68, 191
non-pressurized 68
pressurized 68
with pumps 69
in storage systems 185, 187, 191
for substrate transport 54, 68–9, 96
for vegetable oils 118
plants
   autotrophic 7
   biomass storage 1, 7, 8
   chlorophyll 7
   in climate change 10
   heterotrophic 7
   photosynthesis 1, 3, 7–8, 9
   plant oils see vegetable oils
   water content 25
pollution 13, 23, 25, 38, 39, 62
Portugal 5
post-harvest residues 15, 16
power stations 35, 42
Pro-alcohol Programme (Brazil) 114
pumps
   in anaerobic digesters 54, 68, 69, 70, 86–7, 96
   bellow 69
   centrifugal 69
   circulation pumps 174, 175
   dimensioning 174
   displacement 69
   piping 69
   spiral 69
   substrate pumps 54, 68, 70, 86–7
   vane 69

rapeseed oil 18, 22, 40, 104, 107, 113
renewable energy systems 6, 99, 211, 227, 231, 232–3, 243–7 see also bioenergy projects
research and development 51, 109, 110, 111, 112, 217, 222, 246
residential heating
   central heating systems 121–2, 128, 130, 142–5, 152–69
   dimensioning 121–5, 131, 136–7, 172, 173, 176–7
   hearth appliances 122–5, 128, 130–51, 152
   heat demand 119–28, 129
rotting see decomposition

safety issues
   of anaerobic digesters 76, 77, 90, 96
   of bioenergy projects 233
   of biogas systems 76, 77, 90
   costs 96
   of electricity 76
   fire protection 193
   safety equipment 69, 175
   of stoves 136, 141, 142
sawdust 25, 26, 30, 33, 168
Scandinavia 240–2, 255–7 see also individual countries
solar energy 1, 2, 3, 128, 172
solar radiation 2, 7, 10, 15, 172
solar thermal systems 144, 154, 155, 162–3, 169, 172–4, 247, 248
solid fuel boilers
   air circulation 178
   automatic feed 29, 30, 39, 40, 124, 158–62, 166, 206
   for central heating systems 39, 127, 152–69
   chimneys 156, 160, 164, 178–81
   in CHP systems 16, 44

   combination boilers 168–9
   emissions 39
   heat exchangers 153, 159, 162, 165
   heat output 155, 163, 166, 172, 206
   for large-scale projects 202, 204, 205, 206, 207, 211, 212
   maintenance 156
   manual feed 29, 33, 39, 40, 124
   servicing 157
   soundproofing 164, 178, 181
   specifications 162
   storage tanks 39, 40, 155, 162–3, 172, 173
   see also log-fired central heating boilers; wood boilers; wood pellet boilers; woodchip boilers
solid fuel combustion
   complete 38, 39
   drying of the fuel 36, 38
   flames 37
   gasification process 36, 38
   incomplete 38
   oxidation 37
   process 36–8
   pyrolytic decomposition 36, 37
   size of the fuel 38
   surface area important 38
   temperature requirements 36, 38
   warming of the fuel 36
   wood fires 37
   see also solid fuel boilers
soot 37, 38, 137, 144, 156, 164, 181
South Africa 257
Spain 102, 243, 244
steam
   in CHP systems 42, 47–9
   expansion model engines 48
   steam boilers 48
   steam piston engines 47–8
   steam pressures 48
   steam turbines 48–9
   in wood boilers 170, 181
stem products
   ash content 25, 26, 27, 30
   in bales 26, 34
   calorific value 26
   chemical composition 29
   chlorine content 27
   combustion systems 27
   combustion temperatures 25, 27
   energy content 26
   as an energy source 20, 21
   hay 20, 26
   heating oil equivalent 26
   measurement units 26
   residue 15, 20
   storage 26
   water content 26
   wheat grains 26, 30
   see also straw
storage
   in anaerobic digesters 54, 55, 66, 70–2, 74, 83, 96
   in bales 26, 34
   of biogas 55, 66, 71–2, 83–4, 92, 94
   in bunkers 184, 187–96, 197–200, 206, 207, 209
   conveyor systems 206
   for wood pellets 30, 158, 159, 184–92
   for woodchips 31, 166, 167, 197, 198, 199, 200

filling systems 187, 191, 194, 195, 196
of logs 33, 181–2
of manure 54, 64, 65, 70, 74
pipe systems 185, 187, 191
in silos 184–87, 206
of stem products 26
storage hoppers 138, 139, 140, 161, 162, 184, 196, 197
of straw 26
of substrates 70–1
underground 184, 194–5, 196, 200
of wood 25, 181–200, 206, 207
of wood briquettes 33, 34
of wood pellets 182–196, 197
of woodchips 197–200
see also storage tanks; storerooms
storage tanks
in anaerobic digesters 55, 63, 64, 71, 74, 96
in central heating cookers 144
in CHP systems 44–5
in combustion systems 39, 44–5
dimensioning 172, 173
with solar thermal systems 173
in solid fuel boilers 39, 40, 155, 162–3, 172, 173
storerooms
checklists 193, 194
damp protection 183–4, 194
dimensioning 183
electrical installations 164, 193, 194
filling systems 187, 191, 194
fire protection 193
impact protection mats 191, 194
sloping floors 191–2
soundproofing 178
structural requirements 189, 190–1, 194
for wood pellets 164, 183–4, 187, 189, 191, 194
stoves
air circulation 136–7, 138
ash pans 140
automatic feed 138
automatic ignition 139
chimneys 134, 137, 138, 150
clearances 136
condensation 137
corrosion 140
flue gases 137
flue pipes 134
heat inserts 149
heat output 134, 139, 146, 148, 149, 150, 151
hoppers 138, 139, 140
installation 136
maintenance 138, 140–1
pellet stoves 138–42
problems/solutions 138, 141
rust 141
safety issues 136, 141, 142
servicing 140, 141
specifications 135, 140
stove paint 141
structural requirements 136–7, 140
ventilation 136
wood stoves 134–8, 140
see also tiled stoves
straw
ash content 26
in bales 26, 34
calorific value 26
emissions 27

energy content 21, 26
as an energy source 15, 19, 20–1
heating oil equivalent 26
residues 15, 20
water content 26
substrates in anaerobic digesters
biogas yield 59
dry matter content 59, 64, 65, 75
organic matter content 59
pH value 57, 60, 75
piping 54, 68–9, 96
storage 70–1
substrate level indicators 74
temperature 74
types of 59
see also co-substrates; digestate; manure
sulphur 27
sunflower oil 18, 22, 107
support programmes
for anaerobic digestion projects 97–8
for bioenergy projects 6, 112, 114, 234, 243–58
for electricity generation 234
for gasification schemes 227
for renewable energy systems 6, 99, 211, 227, 243–47
Sweden 5, 240, 241, 242, 245, 246, 256
synthesis gases 24, 29, 109–11, 112, 217, 218, 222–25
in CHP systems 217, 224–5

technology
gasification technology 111, 218, 219–23, 227, 229
technical rules 76, 232, 233, 235
see also research and development
temperature sensors 74
tiled stoves
air circulation 149, 150
basic 146–8
chimneys 150
construction 146–8
flue pipes 150
heat exchangers 146, 147
heat output 146, 148, 149, 150, 151
lining 146, 149, 151
operation 151
problems/solutions 151–2
specifications 150
structural requirements 150–1
warm air tiled stoves 148–50, 151
transport
biofuels used 3, 35, 51, 101–6
electricity as fuel 104
emissions 101–2, 103, 104, 105, 110
infrastructure 105
trees
beneficial effect 9, 12, 13
in carbon cycle 12
as CO_2 store 9, 12, 13
in photosynthesis 9, 12
turbines
in CHP systems 42, 48–51
compressor power 50
gas turbines 48, 49, 50
hot air turbines 48, 50
organic Rankine cycle (ORC) turbines 48, 50

United Kingdom 5–6, 236–7, 245, 249–51
USA

bioenergy projects 244
ethanol use 18, 34, 114, 116
information sources 238, 251–3
permits 237–8

vegetable oils
into biodiesel 22, 104, 108
in blends 107–8
in CHP systems 116, 117, 118
in combustion engines 27, 115, 117
contamination 118
energy efficiency 113–4
in food industry 103, 106, 107, 108
as fuel 27, 40, 101, 103, 104, 107–8, 113–14, 115
heating oil equivalent 40
in heating systems 40
market development 113–14
pipes 118
production 103, 107
quality standards 27
in stationary applications 41, 116, 117
storage 118
technical properties 107–8
viscosity 27, 40, 108, 115, 117, 118
see also rapeseed oil; sunflower oil
ventilation 92, 136, 144, 156, 164

water vapour 36, 51 see also steam
wind power 1, 2, 232, 245, 246, 250
wood
ash content 20, 22, 25, 26, 30
bark 19, 20, 32
by-products 19, 20
calorific value 24, 25, 26
chemical composition 29
in combustion process 36, 38
combustion systems 17, 21, 29–30, 31–2, 33
combustion temperatures 25, 27, 36
cubic metre measurements 25, 26
energy content 26, 124
as an energy source 3, 12, 19–20
gasification see gasification
harvesting 14, 15, 19
heat output 37, 38, 39, 124, 125, 134
heating oil equivalent 26
in heating systems 17, 20, 119, 124–5, 128–30, 201–15
from landscape management 21–2
measurement units 25, 26
mechanically prepared 21, 22
offcuts 29
products 13–14, 19, 20
storage 25, 181–200, 206, 207
waste wood 19, 21, 168, 225, 232, 234
water content 24, 25, 26
wood products 29–30 see also types of product
wood residue 15, 17, 19–20, 21–2, 225
see also wood boilers; wood energy projects
wood boilers 29, 30, 33, 39–40, 44, 170, 178, 181
see also log–fired central heating boilers; wood pellet boilers; wood chip boilers
wood briquettes
ash content 26, 33
calorific value 26
compaction 33
energy content 26
energy density 33

in hearth appliances 124
manufacture 33
measurements 26
storage 33, 34
water content 25, 26, 33
wood energy projects
ash disposal issues 202, 205
boiler selection 202, 204, 205, 206, 207, 211, 212
checklists 203–5
combustion types 208
contracting arrangements 215
cooperatives 214–15
dimensioning 208–7
economic considerations 201, 210–11
fuel supply 202, 203, 204, 212
funding 214
leasing and hire purchase 214
legal considerations 201, 212–16
local community involvement 201, 202, 203–4, 205
monitoring 202
organizational issues 212–16
ownership issues 213–16
planning 201–2, 204, 206, 208–12
professional advice 202, 210
regional partnerships 215–16
site selection 201–2
wood storage 206, 207
wood gas 3, 24, 42, 152–3, 222–3, 224, 225
wood-gasifying boilers see log-fired central heating boilers
wood pellet boilers 169
advantages/disadvantages of different types 159, 160, 162
air circulation 158–9, 160, 164
ash pans 160, 165
bottom-feed system 158–9
chimneys 160, 164, 181
fans 164
flue gases 159, 161–2, 164
flue pipes 164
heat exchangers 159, 162, 165
heat output 163
hoppers 161, 162, 196, 197
operation 165–6
retort system 159–60
servicing 166
specifications 162
top-feed system 161–2
wood pellets 29, 31
ash content 26
auger extraction 161, 184–5, 186, 187, 188–92
bonding agents 30
in bunkers 184, 187–196
calorific value 26, 30
compaction 30, 31
conveyor systems 30, 158, 159, 184–92
drying process 30, 31
energy content 26, 30, 124
filling systems 187, 191, 194, 195, 196
in hearth appliances 124
heat output 39, 124, 125, 139, 163
heating oil equivalent 26, 30
hoppers 138, 139, 140, 161, 162, 184, 196, 197
manufacture 30–1
measurements 26, 30
in silos 184–7
in solid fuel boilers 29, 30, 39, 40, 124, 158–69
storage 25, 182–196, 197

　　　　in stoves 138–42
　　　　underground storage 184, 194–5, 196
　　　　vacuum extraction 184, 186–88
　　　　water content 25, 26, 30
　　　　in wood pellet boilers 158–66, 196
　wood shavings 25, 26, 29, 30, 33
　woodchip boilers
　　　　air circulation 167
　　　　ash pans 166
　　　　heat output 166
　　　　operation 167–8
　　　　specifications 167
　　　　structural requirements 167
　woodchips 29
　　　　ash content 26, 32
　　　　auger extraction 199

　　　　in bunkers 166, 197–200
　　　　calorific value 26
　　　　chopping process 31, 32
　　　　in combination boilers 168
　　　　conveyor systems 31, 166, 167, 197, 198, 199, 200
　　　　energy content 26
　　　　heat output 166
　　　　heating oil equivalent 26
　　　　impurities 31, 32
　　　　measurements 26, 31
　　　　in solid fuel boilers 31–2, 39, 40, 124, 166–8
　　　　storage 197–200
　　　　underground storage 200
　　　　water content 26, 31–2
　　　　in woodchip boilers 166–8